FOR THE ROCK RECORD

EDITED BY

Jill S. Schneiderman
and Warren D. Allmon

FOR THE ROCK RECORD

Geologists on Intelligent Design

University of California Press

Berkeley Los Angeles London

University of California Press, one of the most distinguished university presses in the United States, enriches lives around the world by advancing scholarship in the humanities, social sciences, and natural sciences. Its activities are supported by the UC Press Foundation and by philanthropic contributions from individuals and institutions. For more information, visit www.ucpress.edu.

University of California Press
Berkeley and Los Angeles, California

University of California Press, Ltd.
London, England

Library of Congress Cataloging-in-Publication Data

For the rock record : geologists on intelligent design / edited by Jill S. Schneiderman and Warren D. Allmon.

p. cm.

Includes bibliographical references and index.

ISBN 978-0-520-25758-0 (cloth : alk. paper)

ISBN 978-0-520-25759-7 (pbk. : alk. paper)

1. Evolutionary paleobiology. 2. Historical geology. 3. Earth—Origin. 4. Intelligent design (Teleology). 5. Creation. I. Schneiderman, Jill S. II. Allmon, Warren D.

QE721.2.E85F67 2009

551.7—dc22 2008021095

Manufactured in the United States of America

18 17 16 15 14 13 12 11 10 09

10 9 8 7 6 5 4 3 2 1

This book is printed on Natures Book, which contains 30% post-consumer waste and meets the minimum requirements of ANSI/NISO Z39.48–1992 (R 1997) *(Permanence of Paper)*.

For Meg, who's stuck with me; for Tillie and Caleb, our most precious treasures and most important "job"; and for Mom and Dad, who took me to see the rocks in the first place—JSS

For Jennifer and Alexandra—WDA

CONTENTS

Introduction 1
Jill S. Schneiderman and Warren D. Allmon

PART ONE ROCKS AND BONES

1. Charles Darwin Was a Geologist: Inorganic Complexity and the Rock Record 11
Jill S. Schneiderman

2. Creationist Perspectives on Geology 21
Timothy H. Heaton

3. Missing Links Found: Transitional Forms in the Fossil Mammal Record 39
Donald R. Prothero

4. Pigeonholing the "Dino-birds" 59
Allison R. Tumarkin-Deratzian

PART TWO EDUCATION, POLITICS, AND PHILOSOPHY

5. Pangloss, Paley, and the Privileged Planet: Parrying the Wedge Strategy in Earth Science Education 77
Mark Terry

6. It's Not about the Evidence: The Role of Metaphysics in the Debate 93
Charles E. Mitchell

7. The Misguided Attack on Methodological Naturalism 117
Keith B. Miller

8. *On the Origin of Species* and the Limits of Science 141
David W. Goldsmith

PART THREE ON RELIGION

9. Teaching Evolution during the Week and Bible Study on Sunday 163
Patricia H. Kelley

10. The "God Spectrum" and the Uneven Search for a Consistent View of the Natural World 180
Warren D. Allmon

Selected Resources Relevant to Intelligent Design 241

About the Contributors 247

Index 251

Introduction

JILL S. SCHNEIDERMAN
AND WARREN D. ALLMON

> I want everybody to be smart. As smart as they *can* be.
> A world of ignorant people is too dangerous to live in.
>
> GARSON KANIN (1912–1999), *Born Yesterday*

It seems so long ago. In the fall of 1982, we were both new graduate students in the same geology department, and all the talk was about a federal court case which just a year before had pitted young-Earth creationists against scientists (including one from our department), teachers, and clergy from many denominations. In his decision in that case, *McLean v. Arkansas Board of Education*, U.S. District Court judge William R. Overton offered a detailed definition of science as distinct from religion. His argument seemed to us so clear and convincing we assigned it to the undergraduates in the lab sections we both taught in a course called The History of the Earth and of Life. The decision seemed destined to remove from the purview of science classrooms all discussion of the role of a divine creator in historical geology and evolutionary biology.

And yet in the fall of 2005 we found ourselves on the phone with each other, bemoaning yet another court case and worrying for the intellectual future of our field and our country. Once again we scientists had to defend our disciplines against incursions from realms that would deprive curious thinkers of the opportunity to use science to enrich their understanding of the natural world. This latest challenge, calling itself "intelligent design" (ID), seemed a particularly pernicious variant of the creationism we had hoped was banished a quarter-century before. Learning from the defeats of the 1980s, ID wrapped itself even more tightly in the cloak of science and—publicly at least—steered clear of religion. And it seemed to be gaining ground. It was thus with enormous joy and relief that we learned in late December 2005 that rationality and law had prevailed again. Judge John E. Jones III rendered his decision in *Kitzmiller v. Dover Area School District:* "ID is not science and cannot be adjudged a valid, accepted scientific theory. In making this determination, we have addressed the seminal question of whether ID is science. We have concluded that it is not, and moreover that ID cannot uncouple itself from its creationist, and thus religious, antecedents" (2005, pp. 89, 136).

Yet after this legal victory we knew that all was not well. Although resoundingly vanquished in court, ID in some ways seems to have burrowed more deeply into the public's understanding of science than any of creationism's previous incarnations. A 2007 Gallup poll revealed once again that 45 percent of Americans believe that God created human beings pretty much in their present form sometime within the last ten thousand years. It couldn't be a worse time for such beliefs, when the polis can evaluate only with great difficulty the truthfulness of claims about global climate change. By encouraging the confusion between legitimate religious faith and naturalistic science, ID and its kindred threaten the scientific literacy that our society will need for its very survival in the coming century, and the persistence of these ideas brings renewed urgency to the need for public scientific understanding of Earth and begs for outspoken responses from Earth scientists.

Creationists in general, and intelligent design advocates in particular, mostly criticize organic evolution, especially by Darwinian natural selection. This issue may appear at first to be not a particularly geological problem, except insofar as fossils are preserved in rocks and constitute one of the major categories of evidence for evolution. This book, however, is dedicated to the proposition that intelligent design should be a serious concern to *everyone* interested in science. Although it may not seem so, ID is part and parcel of pseudoscientific explanations for numerous geological phenomena—from the caves of Tennessee to global climate change to erosive mud flows from Mount Saint Helens (e.g., Hitt 1996; Rosin 2007). The First Conference on Creation Geology, held at Cedarville, Ohio, in July 2007, and the $27 million Creation Museum that opened in May 2007 in Petersburg, Kentucky, reflect this resurgent interest in the search for not just biological but geological "evidence" for design. So too does the recent publication of books such as *Geology by Design: Interpreting Rocks and Their Catastrophic Record* (2007), *The Earth Will Reel from Its Place: Scientific Confirmation for Bible Predictions of Geological Upheaval* (2006), and *Geology in the Bible: Earth's Evidence for Intelligent Design* (2005).

For the Rock Record: Geologists on Intelligent Design will appeal to geologists, but we hope it will also be read by other professional scientists, policymakers, teachers, school administrators, and general readers interested in both science and the politics of science. Geology solidly shares with evolutionary biology the thorny issue of interpreting a complex past. Because essays in this volume take Earth science as their launching point, this book differs from others that criticize creationism and intelligent design. In this anthology, geologists confront intelligent design creationists *as geologists* for the first time.

Potential geological responses to intelligent design creationism can be seen as falling into two categories. One is a version of what we think should be the general response of *every* branch of modern science to the attacks on basic rationality that most creationist efforts involve. Virtually all of modern creationism, including intelligent design, assaults *all* science, not just evolutionary biology. If evolution—which is driven mostly but not wholly by natural selection and is accepted as the dominant explanation for the form, history, and diversity of life by essentially every knowledgeable scientist in the world—is wrong, then there is likely something fundamentally misguided about most science, from astrophysics to molecular biology. The language of the infamous intelligent design manifesto makes the attack clear; the Wedge Document states that one of the movement's "governing goals" is to "defeat scientific materialism and its destructive moral, cultural, and political legacies" and "to replace materialistic explanations with the theistic understanding that nature and human beings are created by God" (Center for the Renewal of Science and Culture, 1999). If Earth scientists don't stand up beside our evolutionary biologist colleagues, then we leave our field open to similar victimization.

The second category of potential responses of Earth scientists to intelligent design is to react to its specific implications for our discipline. If a supernatural power designed living things, did it design nonliving things? Did the Earth's interior, crust, and surface evolve naturally, according to the material laws of physics and chemistry, or were they "intelligently designed" as well? If not so designed, why not? How do we know whether they were or weren't? All of the aspects of living things that trouble intelligent design advocates, such as their complexity and what seem to be the abundant traces of long and contingent evolutionary change, also apply to the history of the Earth. Although the mineral quartz is always quartz, many—perhaps most—phenomena of geology, from inclusion-riddled and zoned minerals, to layered rocks and plutons, to mountain ranges and subduction zones, bear exactly the same kinds of "senseless signs of history," as Stephen Jay Gould used to call them (pers. comm.), that strongly argue for their being contingent palimpsests rather than purposefully designed creations. Evolutionary biology and historical geology play by the same rules.

In *Conversations on the Plurality of Worlds* (1686), the French natural philosopher Bernard le Bovier de Fontenelle suggested that God must have designed the universe with human beings in mind. This "anthropic principle"—which suggests that the physical universe is the way it is so as to be suitable for human existence—is usually thought of at an interplanetary scale. That is, if the Earth were slightly farther away from the sun we would freeze, and if we were slightly closer we would roast. This argument presaged today's intelligent design movement. Indeed, the anthropic principle could apply to the arrangement of the continents, changes in sea level and climate, the spatial distribution of natural resources, and the geochemistry of fresh water. If ID is a correct explanation for life, despite all the evidence to the contrary, then it would also surely have to explain geology, and therefore render obsolete two centuries of hard-won knowledge.

Arguments made by modern intelligent design creationists also should matter to Earth scientists because they disregard some of the canonical arguments of our field. For example, in the public controversy over ID, some scientists have pointed out that Charles Darwin (who was, after all, originally a geologist) read and enjoyed, but eventually rejected, the Anglican cleric William Paley's "argument from design." In *Natural Theology*, first published in 1802, Paley used the analogy of the watch to argue that its existence implies a watchmaker, or designer. Paley's explanation of tidal fluctuation—so that ships may leave harbor—is consonant with his view of an omniscient designer and would presumably also be acceptable to today's ID creationists. Yet one of Darwin's great triumphs was to turn Paley on his head—to accept his observations but explain them differently, substituting natural selection for Paley's divine watchmaker, thereby making complexity explainable by the material forces of nature and the materialistic reasoning of science. It is hard to see how one can discard this reasoning in one field because of its implications, and then silently accept its application in others.

As is well known, Darwin also read and admired the work of Charles Lyell (1830), the celebrated founder of modern geology. Darwin embraced Lyellian uniformitarianism—in its "methodological" as well as its "substantive" senses, as Gould (1987) called them—and incorporated it into his theory of evolution through natural selection. As Gould noted, however, Lyell's methodological uniformities of *law* and *process* (the assumption that natural laws and processes do not change over time) are propositions that *all* scientists must accept to do science. They can be distinguished from the substantive notions of uniformity of *rate* (the assumption that the rate at which processes occur presently is the same as the rate at which they occurred in the past) and uniformity of *condition* (the assumption that the state of the Earth always has been as it is today). Confounding them posed serious problems for Darwin. Although he, in Thomas Huxley's words, "burdened himself" by embracing uniformity of rate, leading to the notion of an even and gradual pace for the nature of change, he rejected uniformity of condition. Darwin did so because his observations of change through time on Earth compelled him to, and this led him, for example, to theorize about the process of coral atoll formation. Today's proponents of intelligent design essentially seek to rescramble Lyell's four separate uniformities—selectively rejecting the first two and reinserting uniformity of condition—into Earth science. Regardless of whether ID advocates fall into the young-Earth or old-Earth camp, their insistence that certain structures—because of their purported "irreducible complexity" or some other aspect—could not have developed by processes we can observe acting today would set Earth science back to the days before Charles Lyell.

It is not just life but the Earth itself that is exceedingly complex. One need only try to unravel the story entombed in detrital zircons enclosed in gneisses in Greenland and Australia, for example, to realize that there are multiple layers of intricacy and possibility implied by nonliving as well as living things. How can a zircon grain withstand the punishing cycles of burial, heating, uplift, cooling, and erosion so as to retain a signature of its earliest history? One might be tempted to offer the explanation that "God did it." From James Hutton's "paradox of the soil" (to use Steve Gould's phrase), in which the Earth needed to be old enough to renew itself, to the complex geochemical cycling that fractionates different isotopes of elements, to the systematic crystallization of deep-seated magmas that have produced the Earth's richest ore deposits, complexity and change through time are the Earth's rule. If "irreducible complexity" is allowed into biology, geology will not be far behind and the result will not be good.

The central thesis that drives this book is that Earth science is at the fulcrum of evolutionary thinking and Earth science has never been more important than it is today, when we are faced with so many decisions that require us to understand how the Earth works and how we can live in balance with it. Earth scientists, therefore, must give their own responses to the challenge of intelligent design. The group of outstanding Earth scientists and educators who have contributed to this volume present a variety of perspectives, but are unanimous in their conclusion that intelligent design, like essentially all parts of the modern creationist movement in America, is a political and social agenda masquerading as science.

The book is divided into three sections, the first of which examines geologic and paleontological claims made by creationists of all types, including intelligent design creationists. In the first essay, using examples from Hudson River sediments and Alpine metamorphic rocks, Jill Schneiderman articulates the basis for interpreting real geologic cross-sections. Next, in "Creationist Perspectives on Geology," Tim Heaton shows how geologic claims from all types of creationists contrast with secular geology. In the following two essays, paleontologists directly take on the assertions of purveyors of intelligent design regarding evidence from the sedimentary rock and fossil record. Donald Prothero offers a small selection of the remarkable transitional forms preserved in ancient rocks and effectively refutes the claim of creationists that the sedimentary rock and fossil record contains no transitional forms. Allison Tumarkin-Deratzian uses the close relationship between theropod dinosaurs and birds as a framework to discuss how systems of classification profoundly affect our views of the relationships between organisms.

In the book's middle section, geoscientists tackle education, politics, and philosophy. Some essayists take a philosophical approach to demonstrate that scientific knowledge is distinctive and the understanding that results from it grows continuously. The authors together make a strong case that intelligent design creationism is not science as geologists know and use it. Mark Terry explains how ID creationism looms threateningly over his work as an Earth science educator and makes a case for Earth science education rooted in intellectual history. Charles Mitchell argues that scientific and religious accounts of human origins are founded on different philosophical approaches to knowledge. He describes these differences and discusses how creationist metaphysics leads to erroneous criticisms of Earth and evolutionary science. Keith Miller confronts the recent efforts of intelligent design creationists to redefine science, and in the process reveals widely held misunderstandings of the nature and limitations of science. He illustrates how the attempt by intelligent design advocates to incorporate the supernatural into science undermines empirical inquiry in Earth sciences. David Goldsmith next reminds the reader that Darwin's strict adherence to knowable forces was a radical departure from previous modes of studying the Earth's outer envelope. He unmasks contemporary proponents of intelligent design who falsely claim to be on the cutting edge of science but knowingly employ an outdated intellectual paradigm.

The book concludes with essays on geology and religion. Patricia Kelley discusses how she reconciles her research in paleontology with her faith, while Warren Allmon explores the "scientific" views of proponents of intelligent design, along with the religious views of several leading scientists to determine how or whether they square the nonmaterialism that faith requires with the pursuit or use of the results of materialistic science. Allmon has the final word in this book. As a museum director, he is very much on the front line of science education. His essay helps the reader see that the attack on evolution is just one piece of a wider assault on science that we ignore at our own peril.

Reflecting on evolution with special attention to geological science, the essays gathered here show that intelligent design creationism is part of a larger movement that will adversely alter the nature of science by removing materialism and legitimizing supernaturalism as science. To deny the reality of evolution does a disservice to future generations who will need to contend with enormous and rapid changes on our planet.

REFERENCES

Center for the Renewal of Science and Culture. 1999. The Wedge Strategy. www.antievolution.org/features/wedge.html.

Gould, S. J. 1965. Is Uniformitarianism Necessary? *American Journal of Science* 263:223–28.

———. 1987. *Time's Arrow, Time's Cycle: Myth and Metaphor in the Discovery of Geological Time*. Harvard University Press, Cambridge, MA.

Hitt, J. 1996. On Earth as It Is in Heaven: Field Trips with the Apostles of Creation Science. *Harpers* 293:51–61.

Hutton, J. 1788. Theory of the Earth; or, An Investigation of the Laws Observable in the Composition, Dissolution, and Restoration of Land upon the Globe. *Transactions of the Royal Society of Edinburgh* 1(2): 209–304.

Jones, J. E., III. 2005. Memorandum Opinion. *Kitzmiller v. Dover Area School District*. Case 4:04-cv-02688-JEJ. Document 342. U.S. District Court for the Middle District of Pennsylvania, December 20. www.pamd.uscourts.gov/kitzmiller/kitzmiller_342.pdf.

Lyell, C. 1830. *Principles of Geology*. John Murray, London.

Rosin, H. 2007. God's Harvard: A Christian College on a Mission to Save America. *New York Times Book Review* 112(36): 12–14.

PART ONE • ROCKS AND BONES

ONE · Charles Darwin Was a Geologist

Inorganic Complexity and the Rock Record

JILL S. SCHNEIDERMAN

The Earth's fossil record registers the changes in life on this planet over time; similarly, the Earth's rock record preserves complex structures that record the changes in rocks and minerals over time. Consequently, all the aspects of living things that trouble intelligent design (ID) creationists—their complexity and what seem to be the abundant traces of long and contingent evolutionary change—also apply to all the Earth's materials, whether once living or not. Therefore, one could also ask, if a supernatural power designed living things, what about nonliving things? Did the Earth's interior, crust, and surface evolve naturally, according to the material laws of physics and chemistry, or were they "intelligently designed" as well? If not so designed, why not? How do we know whether they were or weren't? Intelligent design creationists currently ask these seemingly hypothetical questions. Though geological phenomena require naturalistic/materialistic historical explanations, ID creationists offer inadequate, ahistorical ones. In this essay, I use both large-scale and small-scale geologic features to demonstrate the strength of historical explanations to understand extraordinarily complex geologic structures.

In *The Design Revolution: Answering the Toughest Questions about Intelligent Design*, intelligent design creationist William Dembski writes, "As a theory of biological origins and development, intelligent design's central claim is that only intelligent causes adequately explain the complex, information-rich structures of biology and that these causes are empirically detectable" (2004, p. 34). However,

Dembski also broadens the purview of intelligent design and states that "intelligent design is the science that studies signs of intelligence" (p. 10). So, it comes as no surprise to this geologist that Dembski opens chapter 1 of his book on intelligent design with a statement not about organisms but about rocks: "Think of Mount Rushmore—what about this rock formation convinces us that it was due to a designing intelligence and not merely to wind and erosion?" In this regard, he is not unlike the sixteenth-century astrologer/astronomer Johannes Kepler, who postulated that craters on the moon were intelligently designed by moon dwellers. Similarly, ID creationists laud publication of Carl Froede's *Geology by Design: Interpreting Rocks and Their Catastrophic Record* (2007) as an "important reference text for home-schoolers" that asks "what of the rocks beneath our feet?" (Goddard 2007).

Intelligent design creationists believe that, in their words, life is irreducibly complex, and therefore could not have evolved on its own. Thus, a creator must have designed life on our planet. Though focusing explicitly on life, the traditional purview of biologists, this assertion of ID creationists extends to inorganic Earth materials and constitutes an indictment of not only biology but geology. It therefore demands particularly geological responses such as the one I offer in this essay.

Over many decades, creationists have battled with and felt threatened by geologists. The arena of contention has been time. Many first-generation creationists insisted that the Earth is not nearly as old as we geologists would have it. This camp of young-Earth creationists still persists, although the numbers of campers have declined. For geological processes to operate as they clearly do, the Earth simply *must* be very old. Abundant evidence based on years of geoscientific inquiry, investigation, and peer review reveals that the Earth is approximately 4.5 billion years old. In the face of this evidence, some young-Earth creationists have morphed into old-Earth creationists, who accept the ancient age of the Earth. These believers have abandoned time as the arena of controversy. Instead, old-Earth creationists insist that life observed today as living creatures and as fossils entombed in rocks is too complex to have developed on its own over time; it must have been designed by a creator. Many of these old-Earth creationists today base their arguments on *complexity*.

Although the bulk of intelligent design creationists are old-Earth creationists, who allow ample time in Earth's history for geological processes to operate, their insistence that aspects of the natural world are too complex to have developed on their own constitutes an indictment of geology. For example, in the Grand Canyon, the icon of geological thought, in which Earth scientists interpret one of the simplest geologic histories in the United States, intelligent design creationists see evidence of a creator (Wilgoren 2005). Far more complex geologies than that of the Grand Canyon have been studied and explained scientifically by Earth scientists. Yet ID creationists resort to faith to understand this geologic feature that, although stunning, is as "simple as cake"—so simple that geoscientists and Earth science teachers alike refer to it as "layer cake geology." That ID creationists see in the rocks of the Grand Canyon what they consider to be legitimate evidence of a creator suggests that we geologists really don't know what we are talking about when it comes to explaining physical landscapes, structures, and phenomena. But naturalistic/materialistic reasoning suffices to explain exceedingly complex geological features and phenomena as well as simple ones like those of the Grand Canyon. One need not employ claims about the actions of an intelligent creator to explain such physical complexity.

Geologists' explanations of complex structures are based on observations so obvious and routine that geoscientists refer to them as *laws:* the law of stratigraphic superposition (in an undisturbed sequence of strata, the oldest strata lie at the bottom and necessarily higher strata are progressively younger); the law of original horizontality (almost all strata are initially nearly horizontal when they form); the law of original lateral continuity (strata have continuous tabular shapes, "pinching out" laterally to a thickness near zero or abutting against the walls of the natural basin in which they formed); the law of cross-cutting relationships (faults and invading igneous rocks are always younger than the faults or rocks that they transect or intrude); and finally, the law of components (a body of rock is younger than another body of rock from which any of its components are derived). Our understanding is bolstered by the principles of uniformity of law (the idea that natural laws do not change over time) and uniformity of process (the idea that the present is the key to the past). Despite the fact that these laws can guide any careful observer to provide naturalistic explanations for many of Earth's processes and the features that arise from them, ID creationists have sought to develop supernatural explanations—those outside the realm of science—for various features of the Earth.

For example, though plate tectonic theory, one of geology's greatest contributions to twentieth-century science, thoroughly explains the geomorphic features of continents and oceans around the globe, creationists have developed a model termed "catastrophic plate tectonics," which allows a compressed time scale and deploys geological processes to provide a mechanism for the biblical flood (Austin et al. 1994). As a means of finding a source of biblical flood waters, creationist articulations about catastrophic plate tectonics misappropriate numerous well-understood concepts within geology, including mantle convection (the creeping motion of the Earth's rocky mantle in response to unstable variations in its density), geomagnetic reversals (changes in the orientation of Earth's magnetic field), and geochemical processes such as evaporation and precipitation (Baumgardner 2003).

To show how one might be tempted to invoke an intelligent creator to explain the existence of complex features of the Earth at both the macroscopic and microscopic scales, I reproduce and interpret images of some complex Earth structures. At first look, each of these images presents an end product that requires a series of events that might seem impossible without the intervention of a creator. Yet, each set of features has a well-documented history confirmed in the course of the normal scientific research that characterizes the field of Earth science.

GEOLOGICAL SECTIONS

Geological sections or cross-sections show the patterns of rocks as exposed on the side of a road cut or on the wall of a trench. When interpreted carefully, they reveal the histories of sequences of rocks at or near the surface of the Earth; that is, one can infer the order of events that produced the section. In fact the essence of geology, going back to the eighteenth-century Scottish Enlightenment and the days of James Hutton, our science's "founder," is to observe rocks "in the field" and allow them to "tell their stories." To interpret the order of events in a geological section using the laws of superposition, original horizontality, lateral continuity, cross-cutting relationships, and components is to "read" the natural history of that portion of the Earth.

The complex section in figure 1.1 shows a view across New York's Hudson River in the vicinity of the George Washington Bridge. In geological parlance, it shows westward tilted strata of the Newark Basin and the Palisades sill with their nonconformable relationship to folded metamorphic rocks of New York City (Berkey 1948).

How could this complex series of contorted, tilted, and gouged rocks have formed? An intelligent design creationist might well summon the mighty hands of a creator to have upended some rocks while having squeezed and consequently bent the hardest among them, the gneiss and schist, with one hand while using the fingers of the other hand to gouge a channel along which the Hudson River now

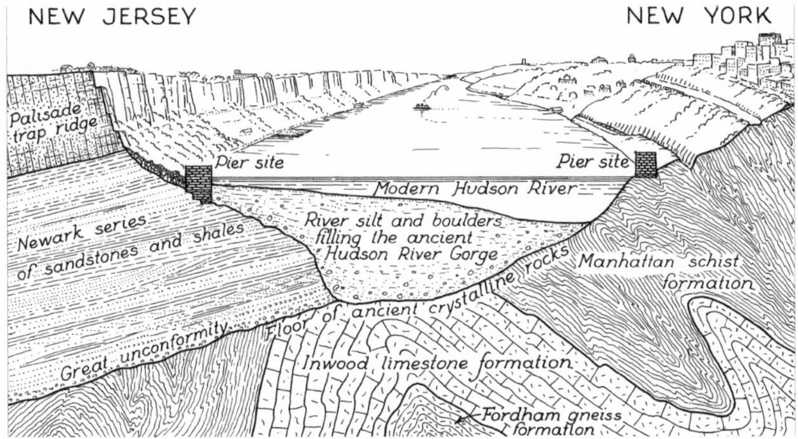

FIGURE I.I
Cross-section across the Hudson River between New York and New Jersey (Berkey 1948).

flows. But to capitulate to such an explanation in the name of "irreducible complexity" would deny the observer the opportunity to understand the natural mechanisms that over time would indeed have produced this section of the Earth.

Features of the Fordham gneiss, Inwood limestone (marble), and Manhattan schist formations, together denoted on the figure as the floor of ancient crystalline rocks, indicate that they originated as ancient bedrock topped by a blanket of sedimentary and volcanic rocks as long as 550 million years ago on the edge of a North American continent rimmed with volcanoes. The ancient geography of this time in Earth history was akin to today's Japanese volcanic islands rimming the coast of China. During a protracted episode of mountain building, known to geologists as an orogeny, the bedrock and volcanic and sedimentary rocks were folded and metamorphosed in a collision that ultimately produced the Appalachian mountain chain.

Experimental and field-based studies indicate that when rocks encounter a change in pressure and temperature as occurs in a zone of collision, they fracture, bend, and generally reorganize themselves so that the mineral grains that originally formed them change their chemical compositions and physical structures. Thus the original rocks metamorphose—change their form—into rocks with new minerals and textures that only barely resemble the protoliths. Indeed this is a creation story, but it is a geological one; the rocks tell their own origin story, one that depends only on natural processes, and we come to understand how the Fordham, Inwood, and Manhattan formations arose.

No intelligent design creationists have thus far directly challenged the veracity of this account of the formation of the basement rocks in southeastern New York State. However, their model of "catastrophic plate tectonics" indirectly condemns such an account. Catastrophic plate tectonics requires "runaway subduction," in which slabs of oceanic crust break off from the Earth's lithosphere and quickly sink deep into the Earth's mantle (Austin et al. 1994). Such rapid tectonics is incompatible with the pace of metamorphism since the recrystallization and deformation necessary for the formation of metamorphic rocks are a slow process. We know this because, for example, though diamond and graphite are both minerals made up wholly of carbon, diamond rings do not transform into graphite in anyone's lifetime; in geological parlance, they persist metastably. Thus it comes as no surprise that intelligent design creationists assert that "the initial state from which the runaway emerged was built into the Earth as God originally formed it" (Baumgardner 2003, p. 12). Nevertheless, we can explain the complexity reflected in the basement rocks of southeastern New York State using only the laws of nature.

The "great unconformity" above the crystalline basement rocks shown in figure 1.1 propels an observer forward in time into the "Newark series of sandstones and shales" and the "Palisades trap ridge." Based on its contained fossils, as well as grain sizes and compositions, we know that this package of rocks is a thick sequence of middle-aged (Mesozoic) sedimentary strata and volcanic layers. Using the laws of original horizontality, lateral continuity, superposition, cross-cutting relations, and components, as well as the principles of uniformity of law and uniformity of process, geologists have been able to discern that the Newark sedimentary strata were deposited in a basin into which seawater never flowed. They became interlayered with igneous rocks that intruded into and erupted on the sedimentary rocks in processes not unlike those that occur today in the rift zones of eastern Africa (Merguerian and Sanders 1994). This sequence of layers formed in association with the opening of the Atlantic Ocean (Olsen 1980). Faults, planes along which rocks have moved against one another, in the area indicate that huge stresses associated with the breakup in the early Mesozoic of a large "supercontinent"—known to geologists as Pangaea—produced a series of basins into which sediments were eroded from adjacent high areas. The rusted red color of some of the sedimentary layers indicates to geologists that, much as metal rusts when exposed to air and water, iron in the sediments interacted with oxygen. Thus, the basins were periodically exposed to air when not covered in shallow water. In many places, "fossil raindrops" preserve a record of rain showers falling on moist muds (Passow 1999–2006). The uniformities of law and process provide ample evidence for the basins' origin in erosion caused by precipitation. Evidence from fossils corroborates this interpretation: worms or other burrowing organisms left tracks, as did some of the earliest dinosaurs as well as other large extinct reptiles such as *Clepysaurus (Rutiodon)*. Above these rock units rest the "river silt and boulders filling the ancient Hudson River gorge," carved through the work of water and ice in the relatively recent geological past (Merguerian and Sanders 1990). Thus, it is by reading the record of the rocks that Earth structures of enormous complexity are explained with the aid of natural laws.

PINWHEEL GARNETS AND MINERAL INCLUSIONS

Photomicrographs of rock samples are photographs of polished, very thin slices of rock taken through a microscope. They frequently reveal that the internal structures of rocks are more complex than is evident to the naked eye. Examined through the microscope, such thin sections are kaleidoscopically beautiful and resemble stained-glass windows. Mineralogists and petrologists (geologists who study the history of the Earth by examining the chemical and physical microstructures of rocks in thin section) determine the histories of rock units by detailing these microstructures and the often multiple generations of events that they conclude must have occurred to produce them.

The image in figure 1.2, whose long dimension does not exceed thirteen millimeters, displays a garnet grain with spiral-shaped inclusion trails (small pieces of one or more types of minerals enclosed in a host mineral) in a rock from the Appalachian mountain range in Vermont. Such garnets have been described as among the most alluring and perplexing microstructures in deformed metamorphic rocks; they look like they have rolled like snowballs in the dirt (Moore 1999). How were they formed? In the face of such tremendous complexity one might be tempted to invoke an artistic and dexterous creator. However, geologists are able to use the law of cross-cutting relationships and the law of components on the microscopic scale for textural analysis, that is, to determine in what order the minerals formed, as well as their knowledge of chemical diffusion, the movement of elements from one part of a solid to another, to outline the remarkable history of the rocks that contain them.

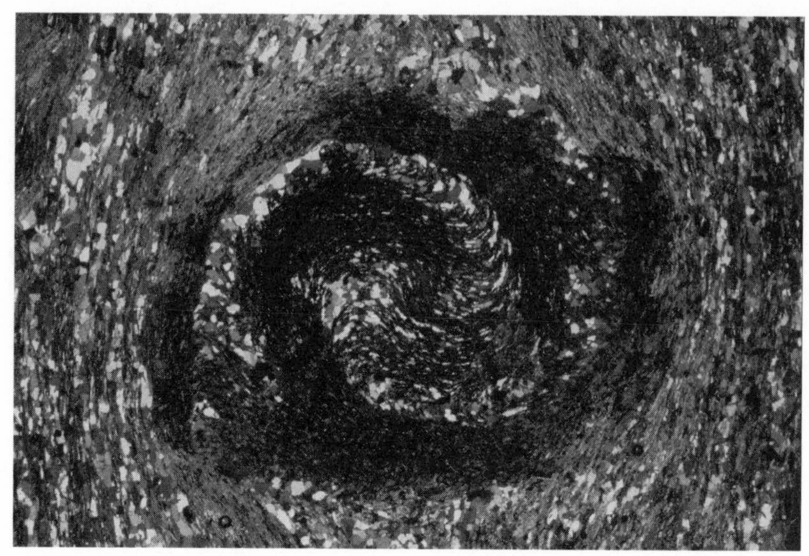

FIGURE 1.2
Garnet grain with spiral-shaped inclusion trails, from the Appalachian Mountains, Vermont (Moore 1999).

The garnet grain in this image contains curved trails of inclusions—incorporated bits of other types of minerals such as quartz and ilmenite—that look like preexisting berries included in pancakes. The origin of these inclusions has been the subject of debate among metamorphic petrologists. Some investigators believe that the mineral inclusion trails indicate that such garnet grains rotated as they grew, while others suggest that the enclosing grains rotated around the garnet crystals. The distinction is not critical because under either interpretation the patterns of inclusions suggest that the garnet grains were not created as we see them; they have a history, and clearly grew as the rocks were actively deformed (Bell 1985; Bell and Johnson 1989; Rosenfeld 1970; Rosenfeld 1987; Schoneveld 1977).

Such curved inclusion trails are a common feature of large crystals, known as porphyroblasts, in rocks from deformed metamorphic regions. They have been a source of intrigue for almost a century and, although complex, have the potential to aid understanding about metamorphic and structural processes that occur during the formation of mountains. Metamorphic petrologists and structural geologists strive to understand the metamorphic and deformation history that a rock has experienced. One problem that such geologists encounter is limited access to information about this history. To determine the early history of a deformed rock, metamorphic petrologists and structural geologists must find "windows" as a way to look into the past.

What of the rocks beneath our feet? They are the result of physical processes that follow natural laws. As the fossil record serves as a window for paleontologists, inclusion-riddled porphyroblasts serve as windows for metamorphic petrologists and structural geologists as they enumerate the sequential development of metamorphic minerals to comprehend episodes of mountain building in the Earth's history (Johnson 1999). Similarly, geologic cross-sections serve as windows for stratigraphers and sedimentologists as they articulate the cycles of deposition, erosion, and uplift recorded in remnants of rock.

REFERENCES

Austin, S. A. 1994. *Grand Canyon: Monument to Catastrophe*. Institute for Creation Research, El Cajon, CA.

Austin, S. A., J. R. Baumgardner, D. R. Humphreys, A. A. Snelling, L. Vardiman, and K. P. Wise. 1994. Catastrophic Plate Tectonics: A Global Flood Model of Earth History. Pp. 609–21 in R. E. Walsh, ed., *Proceedings of the Third International Conference on Creationism*. Creation Science Fellowship, Pittsburgh, PA.

Baumgardner, J. R. 2003. Catastrophic Plate Tectonics: The Physics behind the Genesis Flood. Pp. 113–26 in R. L. Ivey, Jr., ed., *Proceedings of the Fifth International Conference on Creationism*. Creation Science Fellowship, Pittsburgh, PA.

Bell, T. H. 1985. Deformation Partitioning and Porphyroblast Rotation in Metamorphic Rocks: A Radical Re-Interpretation. *J. Metamorph. Geol.* 3:109–18.

Bell, T. H., and S. E. Johnson. 1989. Porphyroblast Inclusion Trails: The Key to Orogenesis. *J. Metamorph. Geol.* 7:279–310.

Berkey, C. P. 1948. Engineering Geology in and around New York. Pp. 51–66 in Agnes Creagh, ed., *Guidebook of Excursions*. Geological Society of America, sixty-first annual meeting, New York.

Dembski, W. A. 2004. *The Design Revolution: Answering the Toughest Questions about Intelligent Design*. InterVarsity, Downers Grove, IL.

Goddard, J. 2007. Geology by Design, November 16. www.shelleytherepublican/category/education/science/junk-science.

Humphreys, D. R. 1990. Physical Mechanism for Reversals of the Earth's Magnetic Field during the Flood. Pp. 130–37 in R. E. Walsh and C. L. Brooks, eds., *Proceedings of the Second International Conference on Creationism*, vol. 2. Creation Science Fellowship, Pittsburgh, PA.

Johnson, S. E. 1999. Porphyroblast Microstructures: A Review of Current and Future Trends. *American Mineralogist* 84:1711–26.

Merguerian, C. 2002. Hofstra University Field Guidebook: A Geological Transect from New York City to New Jersey. www.dukelabs.com/Abstracts%20and%20Papers/1CManual0209.htm.

Merguerian, C., and J. E. Sanders. 1990. *Trip 14: Geology and History of the Hudson River Valley, 28 October 1990*. New York Academy of Sciences Section of Geological Sciences Trips on the Rocks Guidebook, New York.

———. 1994. Post-Newark Folds and -Faults: Implications for the Geologic History of the Newark Basin. Pp. 57–64 in G. N. Hanson, ed., *Geology of Long Island and Metropolitan New York, 23 April 1994*. State University of New York at Stony Brook, Long Island Geologists, Program with Abstracts, Stony Brook, NY.

Moore, R. 1999. Doubly-Curved, Non-cylindrical Surface. wwwtexdev.ics.mq.edu.au/GeoMath/scans/.

Olsen, P. E. 1980. Triassic and Jurassic Formations of the Newark Basin. Pp. 2–40 in W. Manspeizer, ed., *Field Studies of New Jersey Geology and Guides to Field Trips*. New York State Geological Association, fifty-second annual meeting, Rutgers University, Newark, NJ.

Passow, M. J. 1999–2006. Wandering the Watersheds: Hackensack Meadowlands. *Earth to Class*. Lamont Doherty Earth Observatory, Columbia University. www.Earth2class.org/virtualtour/hackensack/hackensack.php.

Rosenfeld, J. L. 1970. Rotated Garnets in Metamorphic Rocks. *Geol. Soc. Am., Spec. Pap.* 129:105.

———. 1987. Rotated Garnets. Pp. 702–8 in C. K. Seyfer, ed., *Encyclopedia of Structural Geology and Plate Tectonics*. Van Nostrand Reinhold, New York.

Schoneveld, C. 1977. A Study of Some Typical Inclusion Patterns in Strongly Paracrystalline Rotated Garnets. *Tectonophysics* 39:453–71.

Wilgoren, J. 2005. Seeing Creation and Evolution in Grand Canyon. *New York Times*, October 6.

Wise, K. P. 2002. *Faith, Form, and Time: What the Bible Teaches and Science Confirms about Creation and the Age of the Universe*. Broadman and Holman, Nashville, TN.

TWO · Creationist Perspectives on Geology

TIMOTHY H. HEATON

Probably no scientific discipline has been more contentious among creationists than geology. At the time Darwin published his *Origin of Species*, the concept of an old Earth with a complex history had been widely accepted among Christians. The threat that this alternate theory of origins posed to theism did not spill over quickly into geology, and even many antievolution preachers supported a harmony between the book of Genesis and long geological ages. Two prominent, competing reconciliations were popular: that the "days of creation" in Genesis were actually long geological periods (the Day-Age Theory) and that there were multiple creations and destructions of life left unrecorded between the first two verses of Genesis (the Gap Theory). In his books on the history of creationism in America, historian Ronald Numbers (1992, 1998, 2006) documents the transition during the twentieth century from these old-Earth creationists to the remarkable rise of young-Earth creationism.

Today we see a different shift in the geological arguments being used by creationists. Proponents of intelligent design (ID) have dropped the demand for a young Earth and focused solely on evidence for a designer. This move is in part a political strategy, used to avoid the legal failures of young-Earth creationists, but many prominent ID creationists are convinced that the evidence for an old Earth is overwhelming and should be embraced. Even before the rise of the ID movement, a group loosely called Progressive Creationists continued to accept the evidence for an old Earth while remaining skeptical about evolution. One such advocate, Hugh Ross, maintains an active ministry and has only peripherally aligned himself with the ID movement. Still other Christians, called Theistic Evolutionists, have fully accepted the case for evolution and hold views on Earth history that are indistinguishable from those of secular geologists (Van Till 1999; Van Till et al. 1990).

The current mix of creationists has created an identity crisis for the overall creation movement, with members of the various camps working in concert on some projects (Dembski 1998b; Meyer et al. 2003; Moreland 1994) while being opponents on other occasions (Hagopian 2001; Moreland and Reynolds 1999; Ross 2004). This chapter reviews the geologic claims of modern young-Earth creationists, progressive creationists, and ID advocates and shows how they contrast with secular geology and with one another on their geological perspectives.

YOUNG-EARTH CREATIONISM

While most other creationists have accepted secular wisdom about geologic history and the age of the Earth, young-Earth creationists have defied this wisdom and attempted to develop an alternative geology that can be accommodated within the few thousand years recorded in the Bible. In light of biblical consistency this approach to geology makes sense. The book of Genesis outlines a history of the Earth from the creation of the world through the origin and cultural history of humans, while it leaves the origin of species and other aspects of biology more open to interpretation.

George McCready Price, the father of modern young-Earth creationism, argued that geology, with its long ages, provided the strongest arguments for evolution and therefore demanded a reinterpretation. In spite of his lack of scientific training, Price (1902, 1916, 1923, 1935) wrote numerous books on geology and developed a strategy that set the stage for the young-Earth creation movement. He attempted to fit a broad array of geologic events into the catastrophic events described in the Bible—particularly the worldwide Flood in the days of Noah. This same strategy was employed by John Whitcomb and Henry Morris in *The Genesis Flood: The Biblical Record and Its Scientific Implications* (1961), the book that converted much of mainstream Protestantism to the young-Earth view.

The problems involved in interpreting Earth history within a few thousand years are staggering. Nineteenth-century geologists recognized many lines of evidence suggesting that the Earth is ancient. For example, most sedimentary rocks are composed of fine-grained minerals and resemble layers forming today in quiet coastal seas. Most of these mineral grains are the result of slow weathering processes on land and are fed to these seas by rivers of predictable velocity. Other minerals have precipitated from ocean water either inorganically or as skeletons of fossil organisms. Under such conditions it would take many millions of years for the thick sedimentary sequences to have formed. By contrast, sediments derived from catastrophic events tend to be distinctive and rare.

Fossils provide further evidence of a long geologic history. Geologists in the early 1800s recognized that fossil species change rapidly through the layers of sedimentary rock but remain in the same relative order from place to place. For example, dinosaurs are all found in a middle section of the record, while human remains are found only near the very top. Geological eras, periods, and epochs, as well as finer-scale units, were named based on fossil content. Because the majority of fossil animals lived in a shallow marine environment, these distinct fossil assemblages could not represent different environmental communities. The logical explanation was that distinct communities of plants and animals lived on Earth during different geological ages. Humans were latecomers and were absent for the great bulk of Earth history. Prior to Darwin's evolutionary proposal, geologists advocated multiple creations and extinctions of life to explain this detailed record. Evolution offered a simpler explanation and one that accounted for the overall pattern of the fossil order.

Both fossils and sediments are indicative of the environment of their formation, and they show that sea level gradually rose and fell many times during Earth history. Many layers in the sequence suggest long periods of stability where great forests and complex coral reef communities developed. Many layers contain mud cracks and buried soil horizons. All this suggests that the sedimentary record did not form quickly or violently. More infrequent events such as ice ages, volcanic eruptions, and meteorite impacts are also recorded and are easy to recognize. Using these clues geologists have reconstructed Earth history in amazing detail.

The discovery of radioactive clocks added precision to the general view of an ancient Earth. The radioactive decay process remains constant under all but the most extreme conditions, and the parent and daughter isotopes can be measured in even the tiniest mineral grains. Radiometric dating began as a simple method for assigning ages to individual rocks but has since expanded to serve a wide variety of applications, with extensive self-checking mechanisms, that can elucidate the complex thermal histories of rock systems. By dating Earth's oldest rocks, along with lunar samples and meteorites, geologists have concluded that our solar system and Earth are about 4.566 billion years old (Dalrymple 2004).

Young-Earth creationists have reacted in various ways to the geologic evidence for an old Earth. Some have flatly denied its validity (H. Morris 1974; J. Morris 1994; Ham et al. 1990), while others have soberly admitted that the evidence does not currently weigh in their favor (Nelson and Reynolds 1999; Oard 1997; Wise 2002). As Dr. D. Russell Humphreys (2005) frankly states, "There is simply too much geological work to be done in too short a time." Those in the latter category invariably add that the authority of the Bible must take precedence in the search for truth. Given this admission, young-Earth creationism must be ruled nonscientific at its foundation. Nevertheless, young-Earth creationists with impressive scientific credentials have sought to build testable models of Earth history based on a literal reading of the book of Genesis.

The goal of young-Earth creationists is to compress the geologic record into a few thousand years by offering either alternate formative processes or familiar processes occurring at much faster rates. Given the magnitude of the problem, these processes must be extraordinary. Aside from the Earth's initial creation, the Bible describes only one event in catastrophic terms with global geologic implications, and that is the great Flood in the days of Noah. Oddly, to account for the details of Earth history, these young-Earth creationists must attribute events to the Flood that go far beyond the Genesis account. Therefore their models are not really biblical even though they are proposed in defense of the Bible.

While virtually all young-Earth creationists attribute much of the sedimentary record to Noah's Flood, there has been a lack of consensus on the exact boundaries. Oard (1990, 2005) and Baumgardner (2002, 2005) follow Price (1923) and Whitcomb and Morris (1961) in attributing the entire fossiliferous portion of the sequence to the Flood. Another view, advocated by Austin (1994), Austin et al. (1994), and Wise (2002), attributes the Paleozoic and Mesozoic eras to the Flood and the Cenozoic era to post-Flood processes. European creationists such as Bush (2004), Robinson (1996), and Scheven (1990) place only early Paleozoic sediments during the Flood.

How do these creationists attribute the order of the fossil record to a single flood? Price and his followers either denied the existence of the order or attributed it to the differential ability of various organisms to escape the rising flood waters. Creationists with geological training such as Wise (2002, 2003) and Wood and Murray (2003) recognize that this explanation fails to explain species with similar mobility and environmental preference, so they have proposed that a complex set of continent-sized, pre-Flood floating forests with distinct organisms succumbed to the Flood in succession, and thus coexisting biogeographic communities could have been converted into worldwide stratigraphic order. As for the sediments themselves, Austin (1994) failed to explain their origin but argued that they could have been rearranged into their current layers by the Flood. Other stratigraphic features such as evidence of living communities and numerous reversals in sea level have not been accounted for. Progress toward explaining the fossil record in a short time frame is meager at best.

Examples of nonbiblical geological concepts accepted by young-Earth creationists in a sped-up time frame include catastrophic plate tectonics and a rapid post-Flood ice age. In both cases they have accepted only the final cycle of these processes and rejected vast evidence for multiple cycles throughout Earth history. The catastrophic plate tectonics model was developed by Austin et al. (1994), Baumgardner (1987, 1990, 2003), and Wise (2002) to explain the diverse evidence for past and present plate movement, such as mid-ocean ridges, deep sea trenches, mountain belts, and volcanoes. In addition they have speculated that catastrophic subduction of the oceanic crust could have initiated the Flood of Noah by vaporizing seawater and generating massive tsunamis. In spite of this proposed benefit, other young-Earth creationists insist that catastrophic plate tectonics is too much of a concession to secular geology and have argued strongly against it (Froede 2000; Oard 2000, 2002; Reed et al. 2000).

Oard (1990) and Vardiman (1993, 2001) have been the main proponents of a recent ice age occurring as an aftermath of Noah's Flood, and this idea has been widely accepted by young-Earth creationists despite lack of biblical support. To force the glacial record into a young-Earth time frame, Oard (1990, 1997, 2005) dismissed multiple cycles of Pleistocene glaciation as mere glacial surges, as well as attributing evidence of older glaciation events to submarine landslides. As is common with young-Earth creationists addressing geological questions, Oard did no field or lab work of his own. He simply scanned the literature for points that could support a single rapid ice age and ignored contrary data. Consequently his writings display a profound ignorance of glacial deposits and the detailed proxy records of ice age climatic cycles. Given that no glaciation is mentioned in the Bible, it is a puzzle that young-Earth creationists feel compelled to believe in an ice age at all.

A recent development in young-Earth creationism is a more honest and sophisticated treatment of radiometric dating techniques. A group called RATE (Radioisotopes and the Age of the Earth) frankly admitted that the evidence for large-scale decay in Earth history was undeniable. To account for this decay within a few thousand years, the group proposed that decay rates increased to millions of times their current levels for brief periods during Creation Week and Noah's Flood (Humphreys 2005). They also freely admitted that the radiation that such accelerated decay would release should have been sufficient to vaporize the Earth and kill all life! The group devoted much of its effort to accommodating these negative side effects and pursuing geologic oddities, such as excess helium in rocks, that might suggest that the decay had occurred recently. In spite of this forthright approach, the RATE project included a propaganda component. A summary book and a DVD titled "Thousands . . . Not Billions" (DeYoung 2005) claim that the RATE team had made a strong scientific case for a young Earth.

While young-Earth creation models follow biblical constraints far more than other brands of creationism, these limits do not necessitate a greater number of miracles. In fact, one biblical constraint is that God performed his creative work in six days and then rested—all before death was introduced into the world by the sin of Adam and Eve. Since the bulk of the geologic record includes fossils (evidence of death), this drives young-Earth creationists to seek mostly natural explanations for Earth's features and events. The most ironic result is that young-Earth creationists have invoked periods of hyperevolution (another sped-up version of a familiar process) following Adam's sin and Noah's Flood to explain the diversity and character of species (Marsh 1944; Wise 2002; Wood and Murray 2003).

Despite the efforts to build a scientific model of a young Earth, the bulk of the young-Earth literature remains highly polemic and designed to sway the uninformed public. Most attention is devoted to expounding weaknesses (real or perceived) in prevailing secular models in hopes that young-Earth creationism will win by default. Unfortunately, even the most credentialed and honest of the young-Earth creationists frequently fall into this habit. Nevertheless, as a group they have managed to propose an impressive faith-based geology that is far more divergent and detailed than any other creationist proposal. While this alternate geology has failed to win any converts outside literal scriptural traditions, it does provide a useful baseline to compare with the views of other creationists.

PROGRESSIVE CREATION

In opposition to young-Earth creationists are a diverse group of scientifically trained Christians who consider the evidence for an old Earth unimpeachable and therefore do not take the Genesis creation story literally. While the term "old-Earth creationists" is sometimes applied, this group is difficult to characterize because of its diversity. Once a Christian relinquishes a literal reading of Genesis, there are many ways to interpret the Genesis account and many choices of natural and supernatural causes. For example, some entirely reject evolution, while others embrace it as God's mode of creation. Another stumbling block to evaluating this group is that they tend to be open-minded and nonspecific about events in Earth history. When these old-Earth creationists criticize their young-Earth counterparts, they cite mostly secular evidence and authorities to promote those parts of their models that coincide with mainstream science rather than offering their own faith-based model to evaluate.

The creationist I will spotlight is Dr. Hugh Ross (1979, 1989, 1993), astronomer and founder of the Reasons to Believe Christian ministry. He has distinguished himself by offering a detailed model of Earth history that stands in stark contrast to that of the young-Earth creationists. As a result, young-Earth creationists have devoted much effort to refuting him (Davidheiser 1998; Oard, Vardiman, and Wieland 2005; Sarfati 2004; Van Bebber and Taylor 1994), for which Ross likes to play the martyr (2004, pp. 13–20). Ross has associated himself peripherally with the intelligent design movement, but he has also criticized that movement for its lack of a historical model and its failure to formally promote the Christian message (Ross 2002, 2006). A study of Ross's model illustrates the difficulties and arbitrariness of mixing empirical data with scripture.

Ross has authored ten books (several in multiple editions) promoting his model and has an active radio ministry. His perspective stems from his unusual conversion to Christianity: he read the scriptures of various world religions and found only the Bible to be harmonious with the facts of astronomy. He promotes the inerrancy of the Bible as strongly as the young-Earth creationists (while interpreting it differently), and he argues that the Big Bang universe and ancient Earth present some of the strongest evidences for the veracity of the Bible. Ross's perspective can be summed up in three major points:

1. The Big Bang represents a discrete beginning to the universe that required a God to initiate, and several Old Testament verses can be interpreted as referring to an expanding universe.
2. The universe, solar system, and Earth are extremely fine-tuned to sustain human life, such that only a God could have created such fortuitous conditions.
3. Natural selection is completely ineffective in creating new species, so all species were created individually by God (i.e., progressive creationism).

Ross's bias toward astronomy stands out sharply in his books. He shows great enthusiasm for astronomical discoveries and fully accepts the prevailing science, while focusing his attention on the special cosmological conditions that produced a habitable Earth. He also accepts the prevailing astronomical and geological chronology and wields the secular evidence for an old universe as a weapon against young-Earth creationists (Ross 2004). But when it comes to biology Ross flatly rejects evolution and engages in weak, polemic arguments to discredit the theory. Geology receives less attention in Ross's books, but it is the arena where his positive and negative views of science merge in a curious fashion.

To harmonize his old-Earth creationism with Genesis, Ross accepts the Day-Age theory, where each day of creation represents a long geologic interval. Unlike the young-Earth creationists, he accepts death before Adam and thereby accepts the fossil record as a historical record of life with humans appearing very late in the record. So in terms of geological history and geologic dating techniques, Ross is fully in-line with modern geological thinking. But in other ways Ross deviates sharply by invoking God, sometimes in strange, paradoxical ways. The crux of Ross's paradox is theological: what things does God have control over, what creative mechanisms are available to him, and how does he choose to act? Consider the following:

"In the context of providing humanity with the richest possible reserves of fossil hydrocarbons, a fixed period of time had to transpire between the epoch when efficient kerogen producers were dominant on Earth and the appearance of human beings. With too little time, not enough petroleum would have been produced. With too much time, most or all of the petroleum would have degraded into methane" (Ross 2006, p. 140). The implication is that God timed the production of petroleum specifically for human benefit, but he was forced to use slow, naturalistic processes to accomplish it. He did not have the power to create petroleum spontaneously or to speed up the geological process (the mechanisms invoked by young-Earth creationists).

Ross describes a stellar model where the early sun underwent a decrease in luminosity that required divine compensation to keep the Earth habitable. His proposed compensations are "outgassing from volcanic eruptions as they pumped just-right quantities of additional greenhouse gases into the atmosphere" (a naturalistic process) and "species of life that were driven to extinction [being] replaced with new species better suited to the cooler conditions brought on by the sun's declining luminosity" (a supernatural process). Ross concludes, "The number of just-right outcomes converging at the just-right times to compensate for the decreasing brightness of the youthful sun seriously strains naturalistic models" (2006, pp. 132–33). This explanation begs the question of why God couldn't simply adjust the sun to maintain a constant energy output, thereby avoiding the need to compensate with numerous miracles on Earth. Ross seems to think that God is restricted to natural processes in the area of astronomy but is free to engage in spontaneous creation when it comes to biology and geology. The reason for this arbitrary distinction seems to lie in nothing more than Ross's training in astronomy and his love for that particular science.

Ross's perpetual hunt for cases of "fine-tuning" in the universe has led him to some outlandish geological examples. In trying to explain the need for ice ages, Ross (2006, p. 173) claims that "large, fast-moving glaciers predominant during ice ages contributed to the formation of many of Earth's richest ore deposits." In reality the high viscosity of glacial ice makes it one of the poorest geological agents for concentrating minerals. Ross (2006, p. 171) also links storm intensity to Earth's rotation rate and claims, "Placing humanity on Earth when the rotation rate had slowed to 24 hours meant that the Creator timed the human era to correspond with the ideal hurricane and tornado era in geologic history—another piece of evidence that the timing of humanity's advent was planned rather than accidental." The fossil record provides no evidence for such a preferential rotation rate. If his geological examples are any indication, Ross's search for fine-tuning seems to be an effort in pure fantasy.

Even in the realm of biology Ross proposes a perplexing mix of natural and supernatural events. He presents a pessimistic view of natural selection and suggests that "every species races an evolutionary clock" as accumulated harmful mutations drive them toward extinction. To repair or replace dying species Ross (2006, pp. 141–44) proposes "transformational miracles" for lower creatures and "transcendent miracles" for higher ones: "With respect to emergence of soulish (birds and mammals) and spiritual characters (humans), outright miraculous acts would appear to be required." God appears capable of creating complex life forms but not of imbuing them with the ability to persist or adapt. Instead, "the existence of numerous 'transitional' forms for whales and horses, among other creatures, further suggests that God performed many creative acts rather than just a few along the way." Only for humans does Ross offer a justification for intermediate fossils as an alternative to evolution: "It seems reasonable that God anticipated the negative impact of (post-Fall) human activity on birds and mammals. One possible scenario is that in the time period prior to Adam and Eve's creation God made a sequence of bipedal primate species, each more skillful at hunting than the one before. Birds and mammals would then have developed better behavioral defenses against the future onslaught of humanity" (2004, p. 237).

Ross (2006, p. 143) makes another interesting biological claim in a geological context: "Speciation and extinction remained roughly balanced before the appearance of human beings . . . but once humans arrived, He ceased making new kinds of life and no longer replaced extinct life-forms (once the seventh day, or era, of rest came)." Ross insists that the scientific evidence supports this disequilibrium between the species origins and extinctions. Here Ross takes advantage of the fact that extinction is a discrete event (the death of the last surviving member of a species) while speciation is a gradual process. A simple analogy shows the fallacy of this argument. Imagine a study of males in a certain city over a period of one year. During the year one hundred males died, but there was not a single indisputable case of a boy becoming a man. Could it therefore be concluded that the number of men is decreasing? Manhood, like speciation, is a gradual and subjective process, so a longer period of time would be required to determine whether the number of men is in equilibrium. In reality there are thousands of cases where biologists disagree over whether two populations represent different species or varieties of the same species. Part of the reason Ross makes this argument may be to deny the relevance of modern biology to the question of species origins.

Young-Earth creationists also take issue with Ross's denial of recent speciation, mostly because they oppose his expanded seventh-day concept. While Ross cites modern astronomical and geological data to refute the short time frame the young-Earth creationists promote, they, in an ironic twist, use the modern evidence for evolution to refute Ross's progressive creationism (Sarfati 2004, p. 236). So both groups utilize scientific data when it suits them while ignoring or dismissing it when it does not.

In his most recent books, Ross (2004, 2006) has offered a long series of "predictions" for the various creationist and naturalistic models so that future discoveries will reveal which one is most accurate. Some geological predictions of Ross's own model are as follows:

> Research increasingly will show that natural "disasters" have struck Earth in a manner that is highly fine-tuned to remove the just-right species at the just-right times to compensate for changes in the solar system and prepare Earth for humanity.

Research increasingly will confirm that Earth's biological history and geological processes were optimally designed to provide humanity with the richest possible fossil fuel deposits.

Research increasingly will confirm that the time interval between some mass extinction events and subsequent mass speciation events is far too brief for any possible naturalistic cause.

While testing hypotheses is an important part of science, vague, subjective predictions such as these offer no meaningful way to evaluate a model, especially when supernatural causes are freely invoked. They merely beg the question of why God would invoke one type of miracle to compensate for a poorly designed feature in some other realm, or why he failed to simply create the necessary resources from scratch.

The way Ross mixes the natural and supernatural makes first-rate comedy. It illustrates that once the door is opened to the supernatural, there are no rules to govern where or how miracles are applied. It is impossible to do science under such conditions. While Ross is to be commended for proposing a detailed model to evaluate, his effort also illustrates why most progressive creationists are reluctant to offer specifics on how God may have intervened in Earth history. Progressive creationism is nothing but an arbitrary blend of science and religion.

INTELLIGENT DESIGN

The intelligent design movement shares many characteristics with progressive creationism, and the two contrast in similar ways with young-Earth creationism. Most ID authors accept the antiquity of the Earth and the details of modern geologic history (see Behe 2007; Dembski 2007; Meyer 2004a, 2004b), but they disparage natural processes as explanations for biological innovations within that history. Instead they appeal to an undefined intelligent designer to explain each complex biological structure. Most of the examples offered are biochemical structures that have left no fossil record of their origin (Behe 1996, 2004; Dembski 1998a, 2002, 2004). Very little attention is paid to geology in ID publications, and this may be because ID proponents have unwittingly selected examples lacking a fossil history in their search for "gaps" in structural development.

Unlike Ross and some young-Earth creationists, the ID advocates have offered no context for the miraculous works of their intelligent designer, and they have produced no historical model to compare with the prevailing scientific model or the young-Earth creation model. This lack of specifics has drawn criticism from all quarters, but it has helped the ID movement attract participants with diverse beliefs and avoid religious labels that would serve as legal liabilities. As a result, ID is largely a negative movement, resembling efforts by young-Earth creationists such as Henry Morris (1974) to bolster creationism simply by attacking evolution. This approach is especially evident in the works of ID writer Jonathan Wells (2000, 2006). While ID advocates have become sophisticated in appealing to a diverse audience using scientific data and statistical probabilities, the strategy is the same: dismiss evolutionary mechanisms and let ID win by default.

Because most ID advocates accept long geological ages and are looking for features that seem to defy natural explanations, geological examples of intelligent design are hard to come by. The only such example that has received much treatment is a unique event in Earth history called the "Cambrian Explosion." This metaphorical name is given to the beginning of the Paleozoic era, when a diverse array of complex animal groups (trilobites, brachiopods, mollusks, etc.) appeared in the geological record over a short period of time, many with no known precursors. The singular nature of this event has long puzzled paleontologists, and for many years it was dismissed as a mere artifact of fossil preservation (Gould 1989). While many other rapid diversifications are documented later in the fossil record, these later diversifications occurred within the basic animal groups that first appeared in the early Cambrian Period.

In proposing the Cambrian Explosion as an example of ID, Meyer et al. (2003) and Meyer (2004a, 2004b) accept the standard geological time scale based on radiometric dating and review the evidence for the rapid appearance of most major animal body plans at the beginning of the Cambrian Period (about 530–525 million years ago). They also quote numerous paleontologists expressing their doubt that conventional natural selection and other known evolutionary mechanisms can account for so much evolution in such a short time. But rather than propose their own theory for this unique event, they employ the usual ID strategy of placing the entire burden on advocates of naturalistic processes: "Thus, for intelligent design to stand as *the best*, rather than just *a plausible*, explanation for the origin of the biological information that arises in the Cambrian, one must show the implausibility of both neo-Darwinian and self-organizational mechanisms as explanations for the origin of the biological information that arises in the Cambrian" (Meyer et al. 2003, p. 368; italics as in original).

In proposing ID as the likely source of the Cambrian animals, Meyer and his colleagues make no attempt to identify the "designer" or his purpose or strategy in this creation. They fail to address obvious philosophical questions, such as: Why would a designer introduce life on a planet, and then wait several billion years before introducing animals with complex organs? Or, why might an intelligent designer need to experiment with diverse body plans when he should know beforehand which would succeed and which would fail? The three Meyer articles are devoted exclusively to dismissing the two naturalistic mechanisms previously listed. They propose ID as the obvious alternative explanation for this well-known mystery, but fail to develop it beyond the level of a simple magic trick. There is nothing remotely scientific about this approach. In this example, as in so many others, ID is nothing more than a synonym for "unknown cause."

While some advocates of young-Earth creationism and progressive creation have moved beyond this polemic approach by offering comprehensive historical models to evaluate, the ID writers have refused to do this (Roberts 2004; Ross 2006). Yet they seem ignorant of the hollowness of this approach. Dembski (2004, p. 329), in reviewing the criticism of ID, complains that "so long as some unknown material mechanism might have evolved the structure in question, Intelligent Design is proscribed." What Dembski fails to acknowledge is that without a comprehensive model, his intelligent designer is also an unknown mechanism! It leaves the "structure in question" as unknown as it was before. The study of material mechanisms allows science to progress in solving life's puzzles. Attributing structures to an undefined intelligent designer is to abandon further inquiry in favor of ignorance. Proponents of ID should either offer historical models that incorporate the specific actions and intentions of their proposed designer or admit that their beliefs are purely a matter of faith.

CONCLUSION

It is understandable that people of faith want to harmonize their beliefs with the observations of science. Science, however, is based on testable hypotheses and comprehensible cause-and-effect relationships. Miracles are excluded because they appeal to something unknown and therefore lead nowhere. Modern young-Earth creationists have attempted to overcome this restriction by proposing detailed historical models with some testable predictions to compare with prevailing scientific models. This effort gives their work a certain scientific respectability. The problem is their unwillingness to abandon religious beliefs, such as a young Earth and a worldwide flood, in the face of overwhelming contrary evidence. The very name of the young-Earth creation movement illustrates that some conclusions are nonnegotiable.

ID advocates have avoided this pitfall by keeping their articles of faith so minimal that they are essentially meaningless. They seek scientific respectability by accepting well-documented scientific conclusions and methodologies. The only prediction they make is that the origin of some structures will remain unexplainable, and they solve these puzzles by advocating an undefined supernatural force. Progressive creationists are harder to characterize but generally fall between these two extremes. Both the young-Earth and ID approaches prevent scientific acceptance. To gain any shade of scientific respectability, the advocates of supernatural causes need to realize and admit where their science ends and their faith begins. They need to propose historical models that contain testable elements, and they need to abide by the results of those tests even if it means abandoning elements of their faith. Without these essentials, creationism will remain a purely religious movement.

REFERENCES

Austin, S. A. 1994. *Grand Canyon: Monument to Catastrophe*. Institute for Creation Research, El Cajon, CA.

Austin, S. A., J. R. Baumgardner, D. R. Humphreys, A. A. Snelling, L. Vardiman, and K. P. Wise. 1994. Catastrophic Plate Tectonics: A Global Flood Model of Earth History. Pp. 609–21 in R. E. Walsh, ed., *Proceedings of the Third International Conference on Creationism*. Creation Science Fellowship, Pittsburgh, PA.

Baumgardner, J. R. 1987. Numerical Simulations of the Large-Scale Tectonic Changes Accompanying the Flood. Pp. 17–30 in R. E. Walsh, C. L. Brooks, and R. S. Crowell, eds., *Proceedings of the First International Conference on Creationism*, vol. 2. Creation Science Fellowship, Pittsburgh, PA.

———. 1990. 3-D Finite Element Simulation of the Global Tectonic Changes Accompanying Noah's Flood. Pp. 35–45 in R. E. Walsh and C. L. Brooks, eds., *Proceedings of the Second International Conference on Creationism*, vol. 2. Creation Science Fellowship, Pittsburgh, PA.

———. 2002. Dealing Carefully with the Data. *Journal of Creation (TJ)* 16(1): 68–72.

———. 2003. Catastrophic Plate Tectonics: The Physics behind the Genesis Flood. Pp. 113–26 in R. L. Ivey, Jr., ed., *Proceedings of the Fifth International Conference on Creationism*. Creation Science Fellowship, Pittsburgh, PA.

———. 2005. Carbon 14 Evidence for a Recent Global Flood and a Young Earth. Pp. 587–630 in L. Vardiman, A. A. Snelling, and E. F. Chaffin, eds., *Radioisotopes and the Age of the Earth*, vol. 2. Results of a Young-Earth Creationist Research Initiative. Institute for Creation Research, El Cajon, CA.

Behe, M. J. 1996. *Darwin's Black Box: The Biochemical Challenge to Evolution.* Free Press, New York.

———. 2004. Irreducible Complexity: Obstacle to Darwinian Evolution. Pp. 352–70 in W. A. Dembski and M. Ruse, eds., *Debating Design: From Darwin to DNA.* Cambridge University Press, New York.

———. 2007. *The Edge of Evolution: The Search for the Limits of Darwinism.* Free Press, New York.

Bush, A. 2004. Flood Models and Chronogenealogy. *Creation Ex Nihilo Technical Journal* 18(1): 62–63.

Dalrymple, G. B. 2004. *Ancient Earth, Ancient Skies: The Age of Earth and Its Cosmic Surroundings.* Stanford University Press, Stanford, CA.

Davidheiser, B. 1998. *Creation, Time, and Dr. Hugh Ross.* Self-published, La Mirada, CA.

Dembski, W. A., ed. 1998a. *The Design Inference: Eliminating Chance through Small Probabilities.* Cambridge University Press, Cambridge, UK.

———. 1998b. *Mere Creation: Science, Faith, and Intelligent Design.* InterVarsity, Downers Grove, IL.

———. 2002. *No Free Lunch: Why Specified Complexity Cannot Be Purchased without Intelligence*, Rowman and Littlefield, Lanham, MD.

———. 2004. The Logical Underpinnings of Intelligent Design. Pp. 311–30 in W. A. Dembski and M. Ruse, eds., *Debating Design: From Darwin to DNA.* Cambridge University Press, New York.

———. 2007. Christian Theodicy in Light of Genesis and Modern Science, version 2.3, March 15. www.designinference.com/documents/2006.05.christian_theodicy.pdf.

DeYoung, D. 2005. *Thousands . . . Not Billions: Challenging an Icon of Evolution, Questioning the Age of the Earth.* Master Books, Forest Green, AR.

Froede, C. R., Jr. 2000. Questions Regarding the Wilson Cycle in Plate Tectonics and Catastrophic Plate Tectonics. Pp. 147–60 in J. K. Reed, ed., *Plate Tectonics: A Different View.* Creation Research Society, St. Joseph, MO.

Gould, S. J. 1989. *Wonderful Life: The Burgess Shale and the Nature of History.* W. W. Norton, New York.

Hagopian, D. G., ed. 2001. *The Genesis Debate: Three Views on the Days of Creation.* Crux, Mission Viejo, CA.

Ham, K., J. Sarfati, C. Wieland, and D. Batten. 1990. *The Answers Book: The 20 Most-Asked Questions about Creation, Evolution, and the Book of Genesis Answered*, rev. ed. New Leaf, Green Forest, AR.

Humphreys, D. R. 2005. Young Helium Diffusion Age of Zircons Supports Accelerated Nuclear Decay. Pp. 25–100 in L. Vardiman, A. A. Snelling, and E. F. Chaffin, eds., *Radioisotopes and the Age of the Earth*, vol. 2. Results of a Young-Earth Creationist Research Initiative. Institute for Creation Research, El Cajon, CA.

Marsh, F. L. 1944. *Evolution, Creation, and Science*. Review and Herald, Washington, DC.

Meyer, S. C. 2004a. The Cambrian Information Explosion: Evidence for Intelligent Design. Pp. 371–91 in William A. Dembski and Michael Ruse, eds., *Debating Design: From Darwin to DNA*. Cambridge University Press, New York.

———. 2004b. The Origin of Biological Information and the Higher Taxonomic Categories. *Proceedings of the Biological Society of Washington* 117(2): 213–39.

Meyer, S. C., M. Ross, P. Nelson, and P. Chien. 2003. The Cambrian Explosion: Biology's Big Bang. Pp. 323–402 in J. A. Campbell and S. C. Meyer, eds., *Darwinism, Design, and Public Education*. Michigan State University Press, Lansing.

Moreland, J. P., ed. 1994. *The Creation Hypothesis: Scientific Evidence for an Intelligent Designer*. InterVarsity, Downers Grove, IL.

Moreland, J. P., and J. M. Reynolds, eds. 1999. *Three Views on Creation and Evolution*. Zondervan, Grand Rapids, MI.

Morris, H. M. 1974. *Scientific Creationism*. Creation-Life, San Diego, CA.

Morris, J. D. 1994. *The Young Earth*. Master Books, Green Forest, AR.

Nelson, P., and J. M. Reynolds. 1999. Young Earth Creationism. Pp. 39–75 in J. P. Moreland and J. M. Reynolds, eds., *Three Views on Creation and Evolution*. Zondervan, Grand Rapids, MI.

Numbers, R. L. 1992. *The Creationists: The Evolution of Scientific Creationism*. Alfred A. Knopf, New York.

———. 1998. *Darwinism Comes to America*. Harvard University Press, Cambridge, MA.

———. 2006. *The Creationists: From Scientific Creationism to Intelligent Design*, exp. ed. Harvard University Press, Cambridge, MA.

Oard, M. J. 1990. *An Ice Age Caused by the Genesis Flood*. Institute for Creation Research, El Cajon, CA.

———. 1997. *Ancient Ice Ages or Gigantic Submarine Landslides?* Creation Research Society Monograph Series, no. 5. Creation Research Society Books, St. Joseph, MO.

———. 2000. Subduction Unlikely—Plate Tectonics Improbable. Pp. 93–145 in J. K. Reed, ed., *Plate Tectonics: A Different View*. Creation Research Society, St. Joseph, MO.

———. 2002. Does the Catastrophic Plate Tectonics Model Assume Too Much Uniformitarianism? *Journal of Creation (TJ)* 16(1): 73–77.

———. 2005. *The Frozen Record: Examining the Ice Core History of the Greenland and Antarctic Ice Sheets*. Institute for Creation Research, El Cajon, CA.

Oard, M. J., L. Vardiman, and C. Wieland. 2005. Cold Comfort for Long-Agers: Hugh Ross' Superficial Interpretation of Ice Core Data. *Creation Matters* 10(5): 1–3.

Price, G. M. 1902. *Outlines of Modern Christianity and Modern Science*. Pacific, Oakland, CA.

———. 1916. *Back to the Bible or the New Protestantism*. Review and Herald, Washington, DC.

———. 1923. *The New Geology: A Textbook for Colleges, Normal Schools, and Training Schools; and for the General Reader*. Pacific, Mountain View, CA.

———. 1935. *The Modern Flood Theory of Geology*. Fleming H. Revell, New York.

Reed, J. K., C. B. Bennett, C. R. Froede, Jr., M. J. Oard, and J. Woodmorappe. 2000. An Introduction to Plate Tectonics and Catastrophic Plate Tectonics. Pp. 11–21 in J. K. Reed, ed., *Plate Tectonics: A Different View*. Creation Research Society, St. Joseph, MO.

Roberts, M. 2004. Intelligent Design: Some Geological, Historical, and Theological Questions. Pp. 275–93 in W. A. Dembski and M. Ruse, eds., *Debating Design: From Darwin to DNA*. Cambridge University Press, New York.

Robinson, S. J. 1996. Can Flood Geology Explain the Fossil Record? *Creation Ex Nihilo Technical Journal* 10(1): 32–69.

Ross, H. 1979. *Genesis One: A Scientific Perspective*. Wisemen Productions, Sierra Madre, CA.

———. 1989. *The Fingerprint of God*, 1st ed. Promise, Orange, CA.

———. 1993. *The Creator and the Cosmos*. NavPress, Colorado Springs, CO.

———. 2002. More Than Intelligent Design. *Facts for Faith* 10(3): 64.

———. 2004. *A Matter of Days*. NavPress, Colorado Springs, CO.

———. 2006. *Creation as Science: A Testable Model Approach to End the Creation/Evolution Wars*. NavPress, Colorado Springs, CO.

Sarfati, J. 2004. *Refuting Compromise: A Biblical and Scientific Refutation of "Progressive Creationism" (Billions of Years) as Popularized by Astronomer Hugh Ross*. Master Books, Green Forest, AR.

Scheven, J. 1990. The Flood/Post-Flood Boundary in the Fossil Record. Pp. 247–266 in R. E. Walsh and C. L. Brooks, eds., *Proceedings of the Second International Conference on Creationism*. Creation Science Fellowship, Pittsburgh, PA.

Van Bebber, M., and P. S. Taylor. 1994. *Creation and Time: A Report on the Progressive Creationist Book by Hugh Ross*. Eden Communications, Mesa, AZ.

Van Till, H. J. 1999. The Fully Gifted Creation. Pp. 159–247 in J. P. Moreland and J. M. Reynolds, eds., *Three Views on Creation and Evolution*. Zondervan, Grand Rapids, MI.

Van Till, H. J., R. E. Snow, J. H. Stek, and D. A. Young. 1990. *Portraits of Creation: Biblical and Scientific Perspectives on the World's Formation*. Eerdmans, Grand Rapids, MI.

Vardiman, L. 1993. *Ice Cores and the Age of the Earth*. Institute for Creation Research, El Cajon, CA.

———. 2001. *Climates before and after the Genesis Flood: Numerical Models and Their Implications*. Institute for Creation Research, El Cajon, CA.

Wells, J. 2000. *Icons of Evolution: Science or Myth? Why Much of What We Teach about Evolution Is Wrong*. Regnery, Washington, DC.

———. 2006. *The Politically Incorrect Guide to Darwinism and Intelligent Design*. Regnery, Washington, DC.

Whitcomb, J. C., Jr., and H. M. Morris. 1961. *The Genesis Flood: The Biblical Record and Its Scientific Implications*. Presbyterian and Reformed Publishing, Philadelphia, PA.

Wise, K. P. 2002. *Faith, Form, and Time*. Broadman and Holman, Nashville, TN.

———. 2003. The Pre-Flood Floating Forest: A Study in Paleontological Pattern Recognition. Pp. 371–82 in R. L. Ivey, Jr., ed., *Proceedings of the Fifth International Conference on Creationism*. Creation Science Fellowship, Pittsburgh, PA.

Wood, T. C., and M. J. Murray. 2003. *Understanding the Pattern of Life: Origins and Organization of the Species*. Broadman and Holman, Nashville, TN.

THREE · # Missing Links Found

Transitional Forms in the Fossil Mammal Record

DONALD R. PROTHERO

INTRODUCTION

The books of the intelligent design (ID) creationists are filled with examples and critiques of evolution from a biological or philosophical perspective, but they pointedly avoid discussing the fossil record or its implications. The longest and most widely read ID book (Behe 1996, p. 27) mentions paleontology only in a few paragraphs (focusing mostly on a common misinterpretation of the Cambrian Explosion). Johnson (1991) repeats many traditional creationist misstatements and lies about the fossil record but does not introduce any new arguments or evidence. The rest of the ID books are similarly silent about the fossil record. Jonathan Wells's (2000) *Icons of Evolution* mentions only horse evolution and *Archaeopteryx* and ignores the rest of the fossil record. The ID creationist high school textbook *Of Pandas and People* (Davis and Kenyon 2004) discusses fossils in a single chapter of a 170-page book. ID creationists have been quoted on numerous occasions as conceding that microevolution occurs and that the Earth may be millions of years old: differences that distinguish them from the more extreme fundamentalist young-Earth creationists, who believe the Earth is only six thousand years old and who generally will not admit that microevolution occurs.

When one looks at the contributors and critical reviewers of ID textbooks such as *Of Pandas and People* (Davis and Kenyon 2004, p. iii), it is clear why they are almost silent about fossils. Although the ID creationists include a few scientists with backgrounds in biology or chemistry, almost none (with the sole exception of Kurt Wise, a student of Stephen J. Gould at Harvard) earned an advanced degree in paleontology from a recognized, accredited noncreationist institution. To my knowledge, not a single ID creationist has ever published a paper on fossils in the peer-reviewed scientific literature, with the one possible exception found in the obscure *Proceedings of the Biological Society of Washington.* A paper on the Cambrian Explosion was snuck into the journal by an editor (who belonged to another ID organization), despite negative reviews and rejection by the other coeditors (see www.biolsocwash.org/; www.expelledexposed.com/index.php/the-truth/sternberg). The little bit that ID creationists write about the fossil record shows that they have no firsthand training in collecting or interpreting fossils, because they rehash old myths and misconceptions from young-Earth creationism literature. As with the young-Earth creationists, their "research" on fossils consists mostly of reading popular books about paleontology and pulling quotes out of context. ID creationists try to impress the uninformed layperson with their Ph.D.'s in biochemistry or physics, but that background has no relevance to understanding paleontology and fossils. Without the appropriate background or training, they are no more qualified to make statements about the fossil record than they are to critique music theory or auto mechanics. Thus, their statements about fossils must always be read with the understanding that they do not actually work on these fossils, and *have probably never even looked at the actual specimens* (nor do they have the training to tell one bone from another if they did).

In a volume such as this, it is useful to examine myths and misconceptions about the fossil record, and give a short update about the truth of these fossils. Since the evolution of birds from dinosaurs is covered elsewhere in this book, I will focus primarily on my area of expertise, fossil mammals.

BUSHES, LADDERS, TRANSITIONAL FORMS, AND "MISSING LINKS"

Much of the public (including most creationists) has mistaken notions about evolution. For example, people sometimes ask, "If humans evolved from apes, why are apes still around?" This question dates back to pre-Darwinian seventeenth- and eighteenth-century notions of life as a "great chain of being" or a "ladder of life" *(scala naturae),* where beings rise from lowly invertebrates to fish to amphibians to reptiles to mammals to humans to cherubim and seraphim and angels and archangels and ultimately to God at the top. But as Darwin and many other scientists have shown, life is not a chain or ladder but a "branching bush," with many ancestral lineages that survive alongside their descendants. When humans evolved from ape-like ancestors, they branched out from a lineage that is still around. Apes did not have to become extinct when some of them evolved into the ancestors of humans.

Closely related to this false notion is the idea that each organism on the "chain of being" is like a "link" in the chain. From this comes the notion of "missing links" that tie together two organisms in the chain. Biologists and paleontologists seldom use this term because of its erroneous connotations that life is a chain of being; however, the public is still confused about this. Despite the creationist denials and misquotations (extensively cited in Prothero 2007), there are hundreds of fossils (and a few living forms) that could be called "missing links" or "transitional forms" between major lineages and species. Yet creationists cannot admit the existence of these forms because they would be conceding that evolution occurs. So they go through all sorts of rhetorical tricks to deny an obvious reality. In some cases, they blatantly deny the truth that is easily demonstrated. When provided with an example of a transitional fossil in a debate, they will ask the evolutionist debater to provide even more transitional forms between that fossil and the fossils that came before and after it (Shermer 2006)! No matter what evidence they are given, their denial mechanisms are so strong that they cannot see what is obvious to any unbiased observer.

Unfortunately, the ID creationists (especially Johnson 1991; Behe 1996; Davis and Kenyon 2004) have borrowed one of the worst habits of the young-Earth creationist authors: quoting scientists out of context. Such a practice of quotation to indicate the opposite of what the author actually meant is a political and rhetorical trick that reflects badly on whoever does it. When the true context of the quote is revealed, it shows that the person who quoted out of context either could not or did not understand what the quote really meant—or that they were intentionally trying to mislead the reader. Davis and Kenyon, in their ID creationist textbook for high school students (2004, p. 96), provide a typical example (borrowed directly from young-Earth creationism books). They quote distinguished paleontologists such as Stephen J. Gould and David Raup to say that the gradual transitions between fossils groups are rare, and that most fossil species are static and unchanged through millions of years. These quotes are from the "punctuated equilibrium" debate that began with Eldredge and Gould (1972). Anyone who bothers to read this subject carefully or read the full context of the quotations will realize that what these paleontologists are saying is that transitional forms are indeed rare, *but they are not unknown*. Contrary to the gradualistic expectations that were widely held prior to 1972, there are good biological reasons for most species to stay stable and unchanging for millions of years, *but nevertheless there are good transitions between many of these species within transforming lineages*. More importantly, we can view each step (different stable species or genera) in a transforming lineage as a transitional form, even though each individual species is relatively unchanging during its time on Earth.

Some creationists (both ID and young-Earth) are aware of the evidence of transforming lineages in the fossil record. They are also literalists about the Noah's Ark story. They must account for all of the millions of life forms on Earth, or else admit that some species have evolved from others since the days of Noah. Creationists claim that Noah took only the created "kinds" (*baramin* in Hebrew) on the boat, and that these "kinds" have since evolved into many more forms (a concession that evolution occurs!). By this method, they claim that there were only about thirty thousand to fifty thousand created "kinds" on board. But then that only gives each "kind" about a cubic meter to live in on the boat—still not much of an improvement on the situation of the animals or the logic of the creationist argument (Moore 1983; McGowan 1984).

This "solution" creates a whole new set of problems. Not only does it concede evolution from the created "kinds," but the "kinds" have no basis in biology at all. When creationist literature is examined, it becomes apparent that sometimes the "kinds" are species, sometimes they are genera, and sometimes they are whole families, orders, or even phyla of animals (Siegler 1978; Ward 1965)! Creationists' arguments are so wildly inconsistent and completely out of line with the known taxonomy of organisms that it is clear that a created "kind" is one of those slippery words that people use to weasel out of difficult spots. As Humpty Dumpty said to Alice (in *Through the Looking Glass*), "Whenever I use a word, it means just what I choose it to mean." Nevertheless, a lot of creationists do "research" that focuses on just this fruitless unscientific version of chasing their own tails, and they even have a name for it: baraminology.

Even when the "baraminologists" claim that the entire evolution of horses or camels is within a single "created kind," they will not admit that there are fossils linking horses to rhinos and tapirs and other perissodactyls, or camels to oromerycids and ultimately to the primitive artiodactyls known as diacodexeids and dichobunids. Thus, even given the huge concession that horse or rhino or camel evolution is real (a concession that most young-Earth creationists deny), they still cannot get around the fact that we have many fossils linking these "baramin" to other "baramin," which completely falsifies their notion that their "kinds" were specially created and did not evolve.

HORSES AND RHINOS AND CAMELS, OH MY!

The fossil record of mammals is full of amazing transitional sequences of fossils, and there are too many to give even a partial list in a short chapter like this. However, some are important to mention because of their extraordinary quality or because they are distorted by the creationists. Davis and Kenyon (2004, pp. 95–96) write that "we cannot form a smooth, unambiguous transitional series linking, let's say, the first small horse to today's horse, land-dwelling mammals to today's whales, fishes to amphibians, or reptiles to mammals." This is a terrible falsehood to put in a high school textbook. They could not have asked for better documented cases than the evolution of horses or whales, or the origin of "reptiles" (meaning amniotes) and "amphibians" (meaning tetrapods). For reasons of space, we will not discuss the latter two here, but they are extensively discussed by Prothero (2007). Davis and Kenyon (2004, p. 96) deny the existence of the evolutionary sequence of horses, yet make no further mention of it anywhere else in their book. Wells (2000, pp. 195–207) discusses how the concepts of horse evolution have moved away from the old linear, straight-line notions to the modern complex phylogeny, but nowhere does he dispute the reality of horse evolution. Instead, his convoluted argument seems to suggest that if more and better fossils force us to change our notions from a simplistic linear model to a more complex, bushy model, we are denying that horse evolution occurred!

The evolution of horses was one of the first transitional series documented after Darwin's book was published. The first studies were published by Thomas Henry Huxley and O. C. Marsh in the 1870s and 1880s, and this example still stands today as one the best transitions we have. As early as 1870, we had fossils of early Eocene (55 million years old) horses such as *Protorohippus* (once called "*Eohippus*" or "*Hyracotherium*"), which were the size of small dogs, had four fingers on their hands and three toes on their feet, and primitive low-crowned teeth. As nearly every textbook in evolution and biology shows, from these simple primitive ancestors, horses went through an amazing sequence of changes. Their side toes were reduced until modern horses run only on the middle digit; their legs got longer for fast running; their teeth became more and more high crowned for eating gritty grasses; their body and brain size increased, their snout became elongated, and their overall skull and body proportions changed dramatically until they resembled horses that we know today. All of those transitional horse fossils are real and well documented. I have personally published research on the *Mesohippus-Miohippus* part of the sequence (Prothero and Shubin 1989; Prothero 1994), and I have collected, identified, and studied horses from many parts of the sequence.

Some parts of this story have been changed and modified as more fossils have been discovered. For example, many of the early renditions of horse evolution were necessarily oversimplifications that showed a simple linear trend in these anatomical changes through time. But we have known for over a century that horse evolution, like that of nearly every other family of organisms on Earth, is bushy and branching, with multiple lineages overlapping in time. Not only do we have this well-documented transformation within the horse lineage, but in recent years we have discovered horses' primitive ancestors. Found in Mongolian rocks, a 58-million-year-old fossil known as *Radinskya* links horses and their close relatives, the tapirs and rhinos, to all the other lineages of hoofed mammals. I have spent much time working with early Eocene horses such as *Protorohippus* and its close relative, the earliest tapir-rhino relative, *Homogalax*. In most features, their teeth and skeletons are nearly indistinguishable, yet there are subtle differences in the cusps and crests of teeth that show that one of them was ancestral to horses, and the other to tapirs and rhinos. Even for the creationists who admit that horses evolved within their own "kind," this link between the "horse kind" and "tapir kind" and "rhino kind" refutes their assertion that there are no macroevolutionary links between "kinds."

The ID creationist Web sites follow the young-Earth creationist model when faced with this reality check: they use quotes out of context. Usually they cite very outdated references about specific details of horse evolution, and leave out just enough information to give the complete opposite impression about what the original text means—to deliberately deceive their reader. Most of these quotations concern the replacement of the old, oversimplified straight-line evolution model with our more modern, bushy branching model. None of the quotations deny that horse evolution occurs, only that it is more complex than originally thought and that we have a much *better* fossil record of horses now, not a worse one. No ID creationists seem to read the more recent literature or deal with new transitional fossils such as *Radinskya*.

In fact, if creationists spent any time at all looking at real fossils, they would be amazed by how subtle the transition is from the primitive relatives of horses, such as *Radinskya* and the phenacodontids, to the early perissodactyls (the odd-toed hoofed mammals, including horses, rhinos, and tapirs). Even more surprising, the earliest Eocene horses, rhinos, and tapirs are very hard to tell apart—yet horses look nothing like the living tapirs or rhinos today. This fact struck me when I was working on my undergraduate research project on early Eocene mammals from the Bighorn Basin of Wyoming. Although the literature on the subject was clear, it was a major challenge trying to tell the earliest horse teeth from the teeth of *Homogalax*, the earliest member of the rhino-tapir lineage. The teeth are virtually identical in size and in cusp-by-cusp detail except that *Homogalax* tends to have slightly better connections of the crests between the cusps (see figure 3.1). The same is true of the skulls and skeletons. All of the early perissodactyl ancestors (horses, rhinos, tapirs, and brontotheres) look so similar when they begin their evolution that only a trained eye can tell them apart. Yet we can trace the evolution of each of these distinct lineages through time, and they soon begin to look very different, so that by the late Eocene, they are dramatically distinct in size and body shape, and even a schoolchild could distinguish between them. This is one of the best examples of how we can document the origin of many modern distinct lineages back to ancestors that converge to the point of being virtually indistinguishable, and how the "kinds" all merge into common ancestors when followed back in time.

But if the evolution of horses is not convincing enough, let's look at my favorite group, the rhinos. They have just as long and dense and detailed a fossil record as horses, yet they have received almost no attention, because their systematics was a mess for decades and nothing could be concluded until the valid species were determined using new collections (Prothero 2005). Once that was done, however, a highly bushy, branching family tree of rhinos results in North America (and a similar pattern in Eurasia), with many different families, species, and genera spanning almost 50 million years. The earliest relatives of rhinos were the early Eocene forms known as *Homogalax*, which also gave rise to tapirs, and yet *Homogalax* is virtually impossible to distinguish from the early Eocene horses. By the middle Eocene we see the split between the tapiroid lineage and the lineages that lead to the three main families of rhinocerotoids. Unlike horses, which evolved mainly in North America with occasional emigrations to Eurasia, rhinos evolved on both hemispheres, and freely migrated back and forth, so their family tree is much more bushy and dominated by sudden migration events than that of the horse. Although most of the species are distinct, we can still see evolutionary trends, particularly in the front of the skull, where the primitive forms have many incisors and small

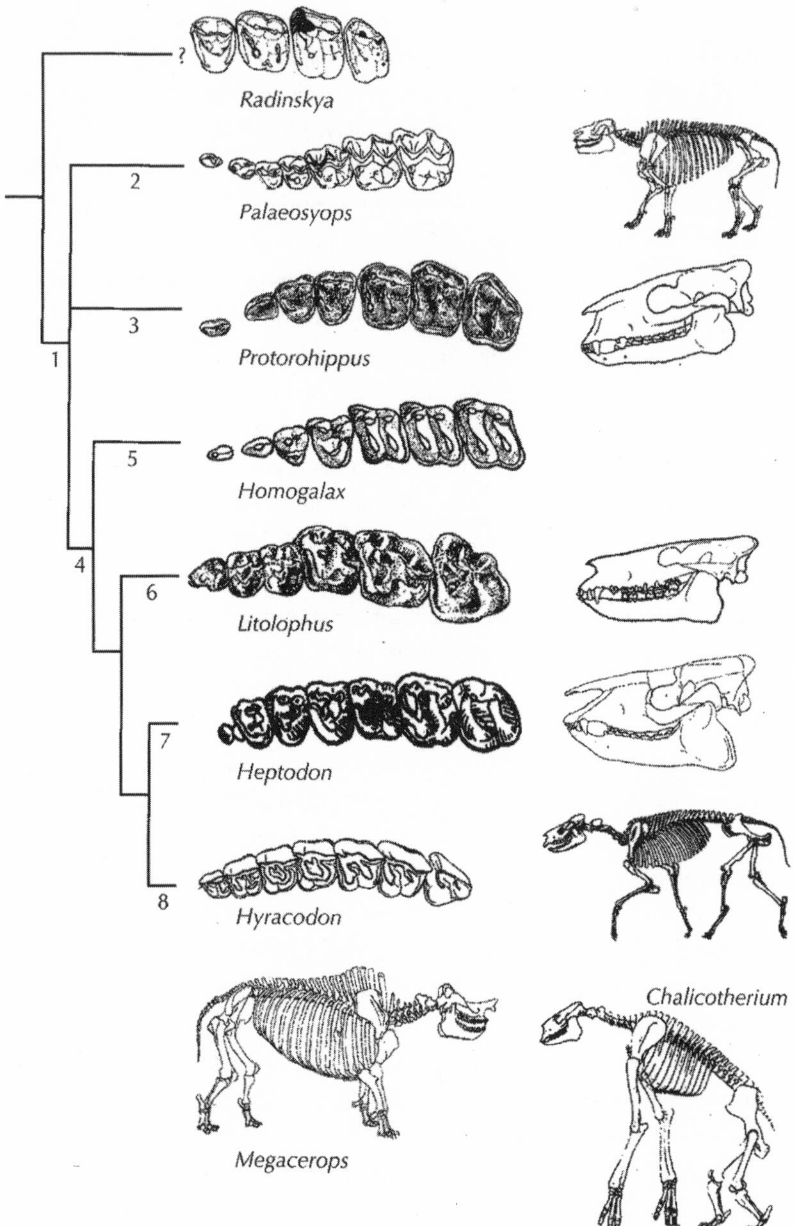

FIGURE 3.1
Radiation of early Eocene perissodactyls (modified from Kemp 2005).

canines. As rhinos evolved, they lost their canines and most of their incisors, and developed sharp, short tusks between their remaining incisors. Prothero (2005) documents many other changes, both gradual and punctuated, such as the size changes in many lineages, and the gradual development in horns within the genus *Diceratherium*. Like horses, rhinos also got larger and more specialized throughout their evolution. They started out with four toes on the front foot, and reduced it down to three by the middle Eocene—but unlike horses, they remain three-toed even today and never became highly specialized one-toed runners like living horses.

Another group with an excellent fossil record that has not received nearly the attention given to horses is the camels. Most people are surprised to learn that extinct camels did not have humps, or that the camel family evolved in isolation in North America. They escaped from this continent only in the late Cenozoic, and reached South America 3 million years ago to evolve into llamas and guanacos and vicuñas, and Eurasia about 7 million years ago, where they evolved into the African dromedary and the Asian Bactrian camels. After all this success, they vanished from their ancestral North American homeland at the end of the last ice age ten thousand years ago. Fossil camels are also surprising in their amazing array of ecological types, far exceeding the limited forms we see today (Honey et al. 1998). The earliest camels were tiny rabbit-sized creatures *(Poebrodon)* that are known from isolated teeth and jaws from the late middle Eocene of Utah, Texas, and California. But by the late Eocene and early Oligocene they had evolved into sheep-sized creatures known as *Poebrotherium* (Prothero 1996), which were common in the Big Badlands of South Dakota. *Poebrotherium* has all the hallmarks of a typical camel: very high-crowned selenodont teeth, long limbs that were nearly fused into a cannon bone, and distinctive features of the skull and skeleton as well. Yet its proportions looked more like those of an antelope or a gazelle, and it apparently had no hump, either. In the late Oligocene and early Miocene, camels underwent an explosive evolutionary radiation into relatively short-limbed varieties (protolabines and miolabines), tiny delicate gazelle-like forms with extraordinarily high-crowned teeth (stenomylines), long-legged, long-necked forms that looked much like the modern guanaco or vicuña (aepycamelines), and even a group which evolved long necks and performed the role of treetop browsers that giraffes occupied in the Old World. Some of these late Miocene and Pliocene "giraffe-camels" were huge as well, with appropriate names like *Gigantocamelus* and *Titanotylopus*. Then, after spreading to Eurasia and South America in the late Miocene and Pliocene, camels dropped in diversity during the ice ages, and only a few species were left when they became extinct on this continent ten thousand years ago.

Yet some "baraminologists" (www.bryancore.org/bsg/Bara99.pdf) essentially concede all of this evolution by shoehorning everything from the tiny rabbit-like *Poebrodon* to the giant giraffe-camels into a "camel kind." That's a pretty incredible stretch for their concept of "baramin," but they still will not concede that there is any link between the "camel kind" and other mammals. But paleontologists have long known that the camels have a closely related group known as oromerycids (Prothero 1998), whose fossils in the early days were mistaken for those of camels until better specimens showed their distinctiveness. The oromerycids, in turn, can be linked right back to the most primitive artiodactyls, the diacodexeids and dicobunids, which are the common ancestor not only of the "camel kind" but also of all the pigs, peccaries, hippos, ruminants, and (as we shall soon see) even the whales.

WALKING WHALES AND MANATEES, SWIMMING ELEPHANTS

Most people are startled enough to learn that most extinct horses had three or more toes on each foot, that most fossil rhinos didn't have horns and that most fossil camels didn't have humps. But they are even more surprised to learn that whales are related to hoofed mammals (ungulates) and are descended from a group of carnivorous ungulates. In debates, creationists love to exploit this public ignorance of the fossil record and zoology by putting up a slide of Bossie the cow and Blowhole the whale, and a ridiculous cartoon of an intermediate between a cow and a whale. But when we said that whales are descendants of ungulates, we did not say cows. Apparently, when creationists hear the words *hoofed mammal*, cows are the only kind they can think of. Actually, hippos would be better models for a modern relative of whales: they are not nearly so different from whales (both are large and aquatic) and the latest genetic evidence puts them as the closest living relative of whales.

Ever since people realized that whales and dolphins were mammals, they have speculated about how they might have evolved from land-dwelling mammals, and from which group of mammals they originated. By the 1830s and 1840s, specimens of huge primitive whales known as archaeocetes were being discovered in the middle Eocene beds of Alabama, but these specimens are fully aquatic, with flippers, tail flukes, and a sinuous body measuring twenty-four meters (eighty feet).

Clearly, the origin of whales must have occurred before the middle Eocene, but nothing was known of their fossil record prior to that time. In 1966, Leigh van Valen and others showed that the skulls and teeth of primitive whales look very much like those of the predatory, archaic hoofed mammals known as mesonychids. Even though mesonychids were land mammals with hooves, there are many similarities in their skull and skeleton (especially their large, serrated, triangular, blade-like teeth) that suggest a close relationship with archaeocete whales. Yet for over a century there were no transitional fossils known between mesonychids and archaeocetes.

Until very recently, paleontologists were comfortable with the idea that whales had arisen from mesonychids, and the fossil evidence seemed to bear this out. Then in the late 1990s, molecular studies showed that among living mammals, the artiodactyls (and particularly the hippos) are most genetically similar to whales (see figure 3.2). This discovery was not too surprising, since artiodactyls and whales are very closely related on the ungulate cladogram, although we always thought they were sister-taxa, not that whales were nested *within* artiodactyls (Prothero, Manning, and Fischer 1988). But in 2001, two independent groups of scientists (Gingerich et al. 2001; Thewissen et al. 2001) found specimens of early whales that preserved the ankle region. Amazing as it seems, these fossils clearly showed that early whales had ankles with the characteristic double-pulley astragalus, the signature feature of the whole order Artiodactyla. Since then, we've rethought the evidence, and now most scientists would agree that whales are a group that evolved from the hippo-anthracothere lineage within artiodactyls, and that mesonychids are the distant relatives of both whales and artiodactyls (Geisler and Uhen 2005).

The best evidence for the origin of whales was found when scientists began to examine the lower Eocene beds of Pakistan. In 1983 Phil Gingerich and colleagues described *Pakicetus*, based on a skull with an archaeocete braincase and teeth intermediate between those of mesonychids and archaeocetes, but lacking ears that were capable of echolocation. *Pakicetus* came from river sediments bordering shallow seaways, suggesting that it might have been a semiaquatic predator that waded in rivers part of the time to find food. The skeleton of *Pakicetus* is quite wolf-like, with long slender limbs and a tail, so it still resembles a mesonychid in most features.

The next development occurred a few years later, when Gingerich, Smith, and Simons (1990) described new specimens of the archaeocete *Basilosaurus* from the middle and upper Eocene deposits of Egypt. Although these new specimens were

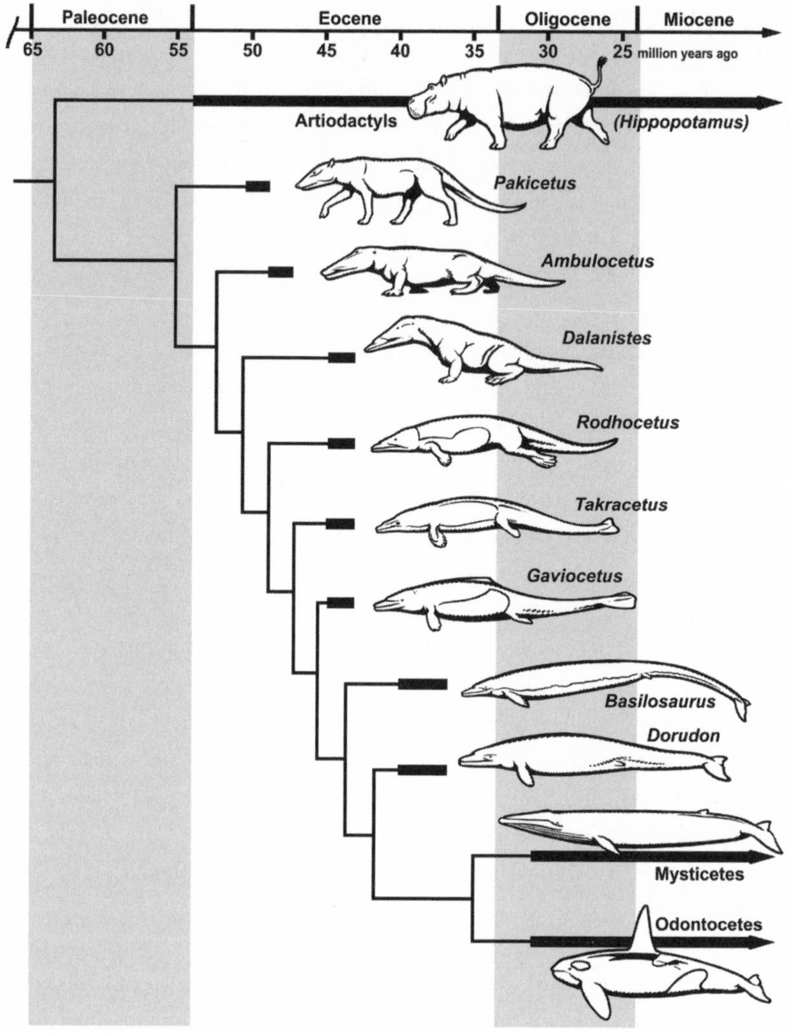

FIGURE 3.2
Whale evolution (courtesy of Carl Buell, pers. comm.).

like other archaeocetes in being fully aquatic, they had something never previously preserved: hind limbs. In most living whales, there are no external hind limbs, but the remnants of the hip and thigh bones are buried in muscles along the spine halfway down the body. *Basilosaurus*, however, had tiny hind limbs that clearly did not function for locomotion: they were about as large as a human arm on a body twenty-four meters long! Like the vestigial hind limbs of modern whales buried inside the body, these tiny limbs can best be explained as functionless relicts of the day when "whales" did walk on land. Since this discovery other archaeocetes, such as *Takracetus* and *Gaviocetus*, have been found to retain vestigial hind limbs.

The crucial discovery occurred when Thewissen, Hussain, and Arif (1994) discovered and described *Ambulocetus natans*, whose name means literally "walking swimming whale." Found in the middle Eocene marine beds of Pakistan, it was about the size of a sea lion, with functional flippers on both its forefeet and huge hind feet (which still had vestigial hooves as well). Its skull and teeth, however, were still like those of mesonychids. Thewissen, Hussain, and Arif suggested that based on its highly flexible vertebrae, *Ambulocetus* swam with an up-and-down flexure of its body similar to the swimming motion of an otter, rather than paddling with its feet like a penguin or seal, or wriggling side to side like a fish. This movement is a precursor to the up-and-down motion of a whale's tail flukes as it swims through the water.

Further discoveries (mostly in the middle Eocene of Pakistan) followed one after another. *Dalanistes*, for example, had fully functional front and hind limbs with webbed feet and a long tail, but was much more whale-like with a longer snout. *Rodhocetus* was more like a dolphin, yet still retained functional hind limbs. As the years go by, more and more transitional whales are being discovered, so that by now the amazing transformation from land mesonychid to whale is one of the best examples of an evolutionary transition in the fossil record (see figure 3.2). This may not make creationists happy, but the fossils cannot be denied.

Creationists are flummoxed by all this new evidence. The ID creationist textbook *Of Pandas and People* (Davis and Kenyon 2004, pp. 101–2) claims "there are no transitional fossils linking land mammals to whales." They could not be more wrong. This false statement is carried over from their 1989 edition into their 2004 edition, yet the 1980s and 1990s yielded an amazing array of transitional whale fossils that clearly link terrestrial land mammals to full-fledged aquatic whales. These fossils have been well documented in many television shows, described in popular books such as Carl Zimmer's (1998) *At the Water's Edge*, and published in high-profile scientific journals such as *Science* and *Nature*, so there is no excuse for creationist ignorance or denial of these fossils. Davis and Kenyon (2004, p. 101) illustrate two extremes of the whale evolutionary sequence (the terrestrial mesonychids and the aquatic archaeocetes) but falsely state that there are no transitional forms between them. Wells (2000) and the other ID creationist books stay away from the evolution of whales entirely, possibly because the case is now so overwhelming that they realize the futility of attacking it.

Elephants and their relatives (the Proboscidea) have an excellent fossil record in the late Oligocene and in more recent rocks, since mastodonts left Africa about 18 million years ago and migrated among all the northern continents. Unfortunately, we are somewhat handicapped because most of their early evolution took place in Africa, and we have a poor fossil record in Africa before the early Oligocene. Nevertheless, we can trace their lineage back from the modern Asian and African elephants and their extinct relatives, the mammoths and mastodonts, through more primitive lineages with a wide variety of tusks and different lengths of trunks. The anancines had two huge, long, straight tusks protruding from their skulls; the stegotetrabelodonts had four long, straight tusks; the deinotheres had two tusks that curled down from their lower jaws; the amebelodonts had their lower tusks flattened into large shovel-like blades. Going back further into the early Oligocene, the famous Fayûm beds of Egypt (home of the archaeocete whales with tiny hind limbs) also produce very primitive, small mastodonts with short jaws and even shorter tusks, known as *Palaeomastodon* and *Phiomia*. The various lineages of proboscideans (elephants, mammoths, and mastodonts) are very primitive and hard to tell apart, typical of the early stages of an evolutionary radiation. These primitive forms can be traced back to the ultimate transitional fossil, *Moeritherium*, from the late Eocene of Egypt. Superficially, *Moeritherium* looked more like a tapir or a pygmy hippo than an elephant, and probably had only a short proboscis, not a long trunk. But a close look at the skull shows that it had very short tusks in the upper and lower jaws, the teeth of a primitive mastodont (*not* those of a tapir or hippo), and the details of the ear region and other parts of the skull (such as the condition of the jugal bones in the zygomatic arch) are unique to the Proboscidea as well. The hippo-like or tapir-like appearance of the skeleton and its preservation in nearshore sediments suggest that *Moeritherium* spent much of its time in water.

All of these fossils have been known for decades, but in the last few years, paleontologists have found even older and better transitional forms. There is the 1984 discovery of an even more primitive proboscidean, *Numidotherium*, from the early Eocene of Algeria (Mahboubi et al. 1984). Although the specimen is very incomplete, it already shows the high forehead, the retracted nasal opening (indicating a short proboscis), short upper tusks, mastodont-like teeth, and the lower front jaw, which was beginning to develop a broad scoop, a diagnostic feature of mastodonts. It was only a meter (three feet) tall at the shoulder, smaller even than

Moeritherium, yet it already had the limb characteristics found in later, larger mastodonts. In 1996 Gheerbrant and others reported the discovery of an even earlier proboscidean, *Phosphatherium*, from the late Paleocene of Morocco. The fossil consists only of an upper jaw (typical of the poor preservation in the Paleocene worldwide), but the teeth already show the distinctive mastodont pattern that appeared at the very beginning of proboscidean evolution. Thus, we now have fossils to trace modern elephants continuously back through many different transitional forms to ancestors that are almost 60 million years old, and that brings us almost to the time when all the hoofed mammal lineages diverged.

Finally, let's look at the manatees, peacefully sleeping away in the shallow warm waters of the tropics and eating sea grasses. According to some historians, the legend of the mermaids may have come from sailors who saw manatees floating upright, feeding their babies at their paired breasts (a configuration also found in humans and elephants), possibly with seaweed draped over them that resembled hair. Close up, of course, they are so plug ugly that they could never be mistaken for beautiful half women/half fish, but never underestimate what months at sea can do for homesick and lonely sailors! This myth, along with the legends of the sirens who tried to lure Odysseus's sailors to their doom with their beauty and seductive songs, is the basis for the name of the order, Sirenia.

When we look at manatees up close, they have many remarkable specializations. Their skulls are unique in several features, especially in the way the upper bones of the skull are modified into a snout. They have horizontal tooth replacement (the same condition as found in their close relatives the Proboscidea), and some have short tusks as well. Their ribs are unique among mammals in that they are extremely dense, heavy, and robust *(pachyostotic)*. These ribs act as diving ballast and help keep the manatee floating at the proper depth. Last but not least, their front limbs are modified into flippers (different in detailed structure from the flippers found in whales or ichthyosaurs), their hind limbs have vanished completely, and their tail is a broad flat horizontal fluke like that found in whales. Its shape is rounded in manatees, but with pointed lobes in dugongs.

Sirenian fossils are well known, although they consist mostly of many broken rib fragments that are dense and heavy, plus a few decent skulls that show their evolution (Domning 1981, 1982). But in 2001, another remarkable transitional form was discovered that clearly catches the sirenians in the act of evolving from land mammals. Known as *Pezosiren portelli* (literally, "Portell's walking sirenian"), it is a nearly complete skeleton from the Eocene of Jamaica (Domning 2001). The skull is much like that of many other primitive sirenians, with all the hallmarks in the skull bones and teeth, and the ribs are thick and heavy, showing that it too was mostly aquatic. But instead of flippers it has four perfectly good walking limbs, with strong shoulder girdles, hip bones, and even well-developed hands and feet! One could not imagine a better transitional form: a creature with all the skull and rib features of manatees, yet still with the ability to walk on land.

In many respects, *Pezosiren* is comparable to *Ambulocetus*, which is a beautiful example of a walking whale making the transition to aquatic life, and to the enaliarctines, which are the transitions from terrestrial bears to seals and other pinnipeds (see Prothero 2007). Just a few years ago, we had no transitional forms to show how terrestrial ancestors of marine mammals such as sirenians, whales, and pinnipeds went back to the sea and became aquatic. Now we have excellent fossil transitions for all three groups! The creationists have written very little about *Pezosiren*, except on their Web sites. There they argue that because it has feet rather than flippers, it simply can't be a sirenian, even though every other part of the skull and skeleton is typically sirenian! As in the case of their dismissal of *Ambulocetus*, they cannot conceive of a transitional form, so any creature that shows these transitional features simply cannot be a whale or a sirenian by their narrow, self-serving definition. This tactic is classic evasion and special pleading, because we couldn't have imagined a better transitional fossil: half manatee, half walking land mammal.

Thus, we have seen that the fossil record of mammals is *full* of transitional forms, showing how nearly all the familiar large ungulates (horses, rhinos, giraffes, elephants, and so on) evolved, and how two groups of marine mammals (whales and sirenians) evolved from land ancestors. We could go on and on about many other excellent transitional series in the mammalian fossil record, but for reasons of space, I instead refer the reader to my recent book *Evolution: What the Fossils Say and Why It Matters* (2007). Now, we need to ask what all this blatant denial of an objective reality tells us about the creationists.

CREATIONISM, DECEPTION, AND INTEGRITY

The way in which ID creationists approach the fossil record raises larger questions about their intellectual integrity and honesty. The ID creationists have made a great effort to deny in public that their movement is religiously motivated, and claim that the "Intelligent Designer" need not be the Judeo-Christian God, but in private they admit that their goals are all about pushing their religious viewpoint (as documented by many authors, e.g., Shermer 2006). They are mostly members of right-wing evangelical Christian churches, and their Discovery Institute in Seattle is funded almost entirely by religious organizations and conservative foundations (see documentation in Shermer 2006). They have tried to hide their religious motivations to get around the separation of church and state enshrined in the U.S. Constitution, but in 2005 they lost badly in the federal trial of *Kitzmiller v. Dover Area School District*. Judge John E. Jones III, a conservative Bush appointee, ruled against the ID creationists and pointed out that their ideas were another thinly disguised effort to inject narrow sectarian religious views into the public school science classroom. In the *Kitzmiller* ruling, he called their ideas "breathtaking inanity." Judge Jones was particularly irritated by the hypocrisy of the ID creationists, who attempt to sound secular when the Constitution is involved, but crow about their religious motives when not in court. "The citizens of the Dover area were poorly served by the members of the board who voted for the intelligent design policy. It is ironic that several of these individuals who so staunchly and proudly touted their religious convictions in public would time and again lie to cover their tracks and disguise the real purpose behind the intelligent design policy" (full text available at www.ncseweb.org).

ID creationism is not about science, but about political power and about dictating the agenda for schools and textbooks now, and eventually exerting control over society. They have described their movement as a "Wedge strategy" to squeeze their religious beliefs (disguised as science) into the public school science classroom. ID creationists play by whatever rules (dishonest if necessary) they need to win. In this chapter, I show how they ignore, distort, or deny evidence; quote out of context; and do many other dishonest things—all in the name of winning their crusade. As someone who was raised attending a Presbyterian church and who learned Bible verses every Sunday, it appalls me to see how unethically these supposed Christian men and women will act in their battle against their perceived foes. It makes one wonder whether they have second thoughts about violating the word and spirit of many parts of the scripture with their lies and deceptions, all to accomplish their goals at the price of their souls.

How do they reconcile this un-Christian behavior with their Christian beliefs? Apparently, to the ID creationists lying and deception are lesser sins than Darwinism, and they are willing to sacrifice their integrity in their crusade against what they believe to be the source of all evils in the world. Their intellectual blinders are so strong that they see only what they want to see, and read only what they want to read, all in the name of their religious beliefs. To ID creationists, pushing their beliefs about the Bible is essential to their religious salvation, and everything else (including science) must be sacrificed so their souls can go to heaven.

ID creationists reveal their true motivations when they speak among themselves. For example, on February 6, 2000, William Dembski told the National Religious Broadcasters, "Intelligent Design opens the whole possibility of us being created in the image of a benevolent God. . . . The job of apologetics is to clear the ground, to clear obstacles that prevent people from coming to the knowledge of Christ. . . . And if there's anything that I think has blocked the growth of Christ as the free reign of the Spirit and people accepting the Scripture and Jesus Christ, it is the Darwinian naturalistic view." At the same conference, Phillip Johnson said, "Christians in the twentieth century have been playing defense. They've been fighting a defensive war to defend what they have, to defend as much of it as they can. It never turns the tide. What we're trying to do is something entirely different. We're trying to go into enemy territory, their very center, and blow up the ammunition dump. What is their ammunition dump in this metaphor? It is their version of creation" (Shermer 2006, p. 109). In 1996 Johnson said, "This isn't really, and never has been, a debate about science. . . . It's about religion and philosophy" (p. 110; interview in *World* magazine, November 30). Jonathan Wells is a follower of the Reverend Sun-Myung Moon and his Unification Church cult (which is vehemently antievolutionary). Wells wrote, "When Father chose me (along with about a dozen other seminary graduates) to enter a Ph.D. program in 1978, I welcomed the opportunity to prepare myself for battle" (2000, p. 110).

Perhaps they should go back to their Bibles, where Proverbs 12:22 states, "Lying lips are an abomination to the Lord."

ACKNOWLEDGMENTS

I thank Jill Schneiderman and Warren Allmon for the invitation to contribute to this volume and for reviewing the chapter. I thank Tim Heaton for his thoughtful comments from both a religious and paleontological perspective. I thank Carl Buell for permission to use figure 3.2. This research was supported by a grant from the Donors of the Petroleum Research Fund of the American Chemical Society, and by NSF grant 03-09538.

REFERENCES

Behe, M. 1996. *Darwin's Black Box: The Biochemical Challenge to Evolution.* Free Press, New York.

Davis, P., and D. Kenyon. 2004. *Of Pandas and People: The Central Question of Biological Origins*, 2nd ed. Haughton Publishing, Dallas, TX.

Domning, D. P. 1981. Sea Cows and Sea Grasses. *Paleobiology* 7:417–20.

———. 1982. Evolution of Manatees: A Speculative History. *Journal of Paleontology* 56:599–619.

———. 2001. The Earliest Known Fully Quadrupedal Sirenian. *Nature* 413: 625–27.

Eldredge, N. and S. J. Gould. 1972. Punctuated Equilibria: An Alternative to Phyletic Gradualism. Pp. 82–115 in T. J. M. Schopf, ed., *Models in Paleobiology*. Freeman Cooper, San Francisco, CA.

Geisler, J. H., and M. D. Uhen. 2005. Phylogenetic Relationships of Extinct Cetartiodactyls: Results of Simultaneous Analyses of Molecular, Morphological, and Stratigraphic Data. *Journal of Mammalian Evolution* 12:145–60.

Gheerbrant, E., J. Sudre, and H. Cappetta. 1996. A Paleocene Proboscidean from Morocco. *Nature* 383:68–70.

Gingerich, P. D., M. U. Haq, I. S. Zalmout, I. H. Khan, and M. S. Malakani. 2001. Origin of Whales from Early Artiodactyls: Hands and Feet of Eocene Protocetidae from Pakistan. *Science* 293:2239–42.

Gingerich, P. D., B. H. Smith, and E. L. Simons. 1990. Hind Limbs of Eocene *Basilosaurus*: Evidence of Feet in Whales. *Science* 249:154–57.

Gingerich, P. D., N. A. Wells, D. E. Russell, and S. M. Ibrahim. 1983. Origin of Whales in Epicontinental Remnant Seas: Evidence from the Early Eocene of Pakistan. *Science* 220:403–6.

Honey, J., J. A. Harrison, D. R. Prothero, and M. S. Stevens. 1998. Camelidae. Pp. 439–62 in C. Janis, K. M. Scott, and L. Jacobs eds., *Evolution of Tertiary Mammals of North America*. Cambridge University Press, Cambridge, UK.

Johnson, P. E. 1991. *Darwin on Trial*. Regnery Gateway, Washington, DC.

Kemp, T. S. 2005. *The Origin and Evolution of Mammals*. Oxford University Press, Oxford, UK.

Mahboubi, M., R. Ameur, J.-Y. Crochet, and J. J. Jaeger. 1984. Earliest Known Proboscidean from the Early Eocene of North-West Africa. *Nature* 308:543–44.

McGowan, C. 1984. *In the Beginning: A Scientist Shows Why the Creationists Are Wrong*. Prometheus Books, Buffalo, NY.

Moore, R. 1983. The Impossible Voyage of Noah's Ark. *Creation/Evolution* 11:1–40.

Prothero, D. R. 1994. Mammalian Evolution. Pp. 238–70 in D. R. Prothero and R. M. Schoch, eds., *Major Features of Vertebrate Evolution*, vol. 7. Paleontological Society Short Courses in Paleontology. Paleontological Society, Pittsburgh, PA.

———. 1996. Camelidae. Pp. 591–633 in D. R. Prothero and R. J. Emry, eds., *The Terrestrial Eocene-Oligocene Transition in North America*. Cambridge University Press, Cambridge, UK.

———. 1998. Oromerycidae. Pp. 426–30 in C. Janis, K. M. Scott, and L. Jacobs, eds., *Evolution of Tertiary Mammals of North America*. Cambridge University Press, Cambridge, UK.

———. 2004. *Bringing Fossils to Life: An Introduction to Paleobiology*, 2nd ed. WCB / McGraw-Hill, New York.

———. 2005. *The Evolution of North American Rhinoceroses*. Cambridge University Press, Cambridge, UK.

———. 2007. *Evolution: What the Fossils Say and Why It Matters*. Columbia University Press, New York.

Prothero, D. R., E. Manning, and M. Fischer. 1988. The Phylogeny of the Ungulates in M. J. Benton, ed., *The Phylogeny and Classification of the Tetrapods*. Repr., *Systematics Assoc. Spec*. 35(2): 201–34, Clarendon, Oxford, UK.

Prothero, D. R., and N. Shubin. 1989. The Evolution of Oligocene Horses. Pp. 142–75 in D. R. Prothero and R. M. Schoch, eds., *The Evolution of Perissodactyls*. Oxford University Press, New York.

Shermer, M. 2006. *Why Darwin Matters: The Case against Intelligent Design*. Times Books, New York.

Siegler, H. R. 1978. A Creationist's Taxonomy. *Creation Research Society Quarterly* 15:36–38.

Thewissen, J. G. M., S. T. Hussain, and M. Arif. 1994. Fossil Evidence for the Origin of Aquatic Locomotion in Archaeocete Whales. *Science* 263:210–12.

Thewissen, J. G. M., E. M. Williams, L. J. Roe, and S. T. Hussain. 2001. Skeletons of a Terrestrial Cetacean and the Relationships of Whales to Artiodactyls. *Nature* 413:277–81.

Van Valen, L. 1968. Monophyly or Diphyly in the Origin of Whales. *Evolution* 22:37–41.

Ward, R. R. 1965. *In the Beginning*. Baker Book House, Grand Rapids, MI.

Wells, J. 2000. *Icons of Evolution: Science or Myth? Why Much of What We Teach about Evolution Is Wrong*. Regnery, Washington, DC.

Zimmer, C. 1998. *At the Water's Edge: Macroevolution and the Transformation of Life*. Free Press, New York.

FOUR · Pigeonholing the "Dino-birds"

ALLISON R. TUMARKIN-DERATZIAN

INTRODUCTION

One of the central claims of the intelligent design movement is that certain features of biological organisms are "irreducibly complex," with such tightly integrated components that removal of any one part renders the system incapable of functioning (Behe 1996). The existence of irreducibly complex structures has been repeatedly put forth as evidence against evolution via natural selection, on the grounds that such a system could not be assembled incrementally over time. Although most irreducible complexity arguments deal with the molecular and cellular levels, such as the vertebrate blood-clotting response or the bacterial flagellum (Behe 1996), a common organismal-level target has been the evolution of feathers and flight in birds. An in-depth refutation of the supposed irreducible complexity of the avian flight apparatus has been outlined by Gishlick (2004), on the grounds that the fossil record preserves evidence of multiple stages in the evolution of feathered flying birds from nonfeathered, earthbound dinosaurs. The purpose of this essay is not to repeat those arguments. Gishlick's case is eloquently presented, but he is largely preaching to the choir. Those who accept the existence of "dino-birds" need little convincing that the avian flight apparatus is not irreducibly complex. On the other hand, those who doubt the very existence of a dinosaur-bird transition will see no logic in the argument *against* irreducible complexity. At the most basic level, a key issue in the public mind has remained the same since the mid-nineteenth century, when *Archaeopteryx* from the Late Jurassic of Germany was first described (Meyer 1861). Are there convincing intermediate forms that link reptiles and birds?

From the time of its initial discovery, *Archaeopteryx* has been regarded as an important player in the story of bird evolution, because it possesses features originally believed unique to birds (e.g., feathers) and other features more characteristic of traditional reptiles (e.g., a long bony tail, teeth, clawed fingers). So similar is *Archaeopteryx* to certain small theropod dinosaurs, that one specimen lacking the famous feather impressions was originally misidentified as the dinosaur *Compsognathus* (Wellnhofer 1974). As fossils of other extinct birds have been discovered, it has become clear that "reptilian" features are not unique to *Archaeopteryx*. All modern birds lack teeth, but several lineages of fossil birds possessed toothed jaws. *Confuciusornis* from the Early Cretaceous of China retains claws on its wings. Until recently, however, there remained one feature that reliably separated *Archaeopteryx* and other early birds from traditional dinosaurs—feathers.

Beginning in the mid-1990s, the dividing line between theropod dinosaur and bird became noticeably blurry. Spectacularly preserved fossils of small theropods from Cretaceous rocks in Liaoning, China, showed traces of feathers or feather-like features. The first "feathered dinosaur" to be described (somewhat ironically, as a bird) was *Sinosauropteryx* (Ji and Ji 1996). Although *Sinosauropteryx* lacks true feathers, it exhibits a filamentous body covering comprising structures that have been widely interpreted as protofeathers. Since then, true feathers similar to those seen in *Archaeopteryx* and modern birds have been described from several other small theropod dinosaurs, such as *Caudipteryx* (Ji et al. 1998) and *Microraptor* (Xu et al. 2003). It is now apparent that feathers can no longer be considered features unique to birds. *Microraptor* even has asymmetrical flight feathers identical to those of modern birds, suggesting that the capacity for flight may have evolved in non-avian theropod dinosaurs.

As discoveries of additional feathered dinosaurs further blur the traditional distinction between small theropods and birds, one might expect increased public acknowledgment of an evolutionary connection between dinosaurs and birds. This, however, has not necessarily been the case. Even as more and more feathered theropods are reported, antievolutionary arguments have continued to focus primarily on *Archaeopteryx*, and the supposed absence of intermediate or transitional forms in the fossil record. Until recently, the feathered theropods of China have been largely discounted as irrelevant to the discussion. This is mostly a result of the age of the Chinese fossils in relationship to the age of *Archaeopteryx*, combined with a widespread misunderstanding of the way in which paleontologists determine relationships between organisms.

Opponents of evolution have repeatedly used four main arguments to contest the relationship of *Archaeopteryx* and the feathered dinosaurs to the evolution of birds. Although originally advanced by young-Earth creationists in the 1970s, similar claims are now being recycled by adherents of the intelligent design movement. These arguments are as follows:

1. The feathered dinosaurs cannot be bird ancestors because they are millions of years younger than *Archaeopteryx* and occur in the same rocks as true birds (e.g., Wells 2002; Robertson 2004).
2. *Archaeopteryx* cannot be considered a part of the dinosaur-bird transition because it is a bird (e.g., Gish 1973; Sarfati 1999).
3. *Archaeopteryx* cannot be considered a part of the dinosaur-bird transition because it is a dinosaur (e.g., Wells 2000).
4. Neither *Archaeopteryx* nor the Chinese feathered dinosaurs are relevant because a transitional form should be intermediate between two groups in all its features, rather than simply having a mosaic of both groups' features (e.g., Morris 1974; Koons 2004).

ARGUMENT 1

The feathered dinosaurs cannot be bird ancestors because they are millions of years younger than Archaeopteryx *and occur in the same rocks as true birds.*

This argument assumes that paleontologists are, in fact, claiming that the known feathered dinosaurs are bird ancestors. This is not the case, but it is easy to make this misinterpretation if one doesn't fully appreciate the tree diagrams commonly used to illustrate the dinosaur-bird connection.

Paleontologists today do not generally use the familiar Linnaean taxonomic system (Kingdom-Phylum-Class-Order-Family-Genus-Species). The reason for this is that the Linnaean system, being essentially a set of nested boxes, is designed for categorical classification only. It is not designed to recognize lineages. (The specific limitations of the Linnaean system tie in more closely with arguments 2 and 3, and will be discussed later.) Most modern paleontologists use a different system—phylogenetic systematics, or cladistics. The advantage of cladistics is that it is far superior to the Linnaean system when trying to work out relationships between organisms on the family tree of life. The disadvantage is that it can be quite confusing to the nonspecialist, especially when the question of fossil ages is involved.

Cladistic methodology works out relationships between organisms based on distributions of "shared derived characters." These are characteristics that organisms share by virtue of their having been inherited from a common ancestor. For example, mammals have mammary glands that produce milk for their offspring. The presence of mammary glands is a shared derived character common to modern mammals because it has been inherited from a common mammalian ancestor that possessed this trait. Most mammals also have four limbs, but rather than being a shared *derived* character of mammals, the presence of four limbs is considered a shared *primitive* trait, because most tetrapods (all vertebrates except fish) also possess four limbs. The presence of four limbs is therefore not a useful character to use when trying to recognize a mammal, because many other nonmammalian vertebrates also have four limbs. Possession of four limbs would be a shared derived character for tetrapods, but a shared primitive character for the more specific group mammals. The absence of four limbs (or more specifically the loss of hind limbs), on the other hand, would be a shared derived character for the subgroup of mammals that includes modern whales and dolphins. The point is that whether a character is considered primitive or derived can vary depending upon where the organism is on the tree. A character that is derived at one level (four limbs for tetrapods) can be primitive at a higher level (four limbs for mammals).

Whether a character is considered derived or primitive at a given level can also change as more fossil organisms are discovered. Such is the case with feathers. As birds are the only modern vertebrates with feathers, it was long believed that the presence of feathers was a shared derived character of birds. The discovery of feathers in nonavian theropod dinosaurs has shown that feathers are actually a shared primitive character of birds, and a shared derived character of a larger group that includes birds and several lineages of small theropods.

Relationships determined using cladistic methodology are depicted on a cladogram, a diagrammatic map of the distributions of shared derived characters (see figure 4.1A). The closer two organisms or groups of organisms are on a cladogram, the more derived characters they share. Cladograms can be easily misinterpreted by nonspecialists because their structure is similar to the more familiar family tree, on which the lower branches represent progressively older generations. Although it possesses a similar branching structure, a cladogram is not a genealogy depicting ancestor-descendant relationships; it is simply a map of characters.

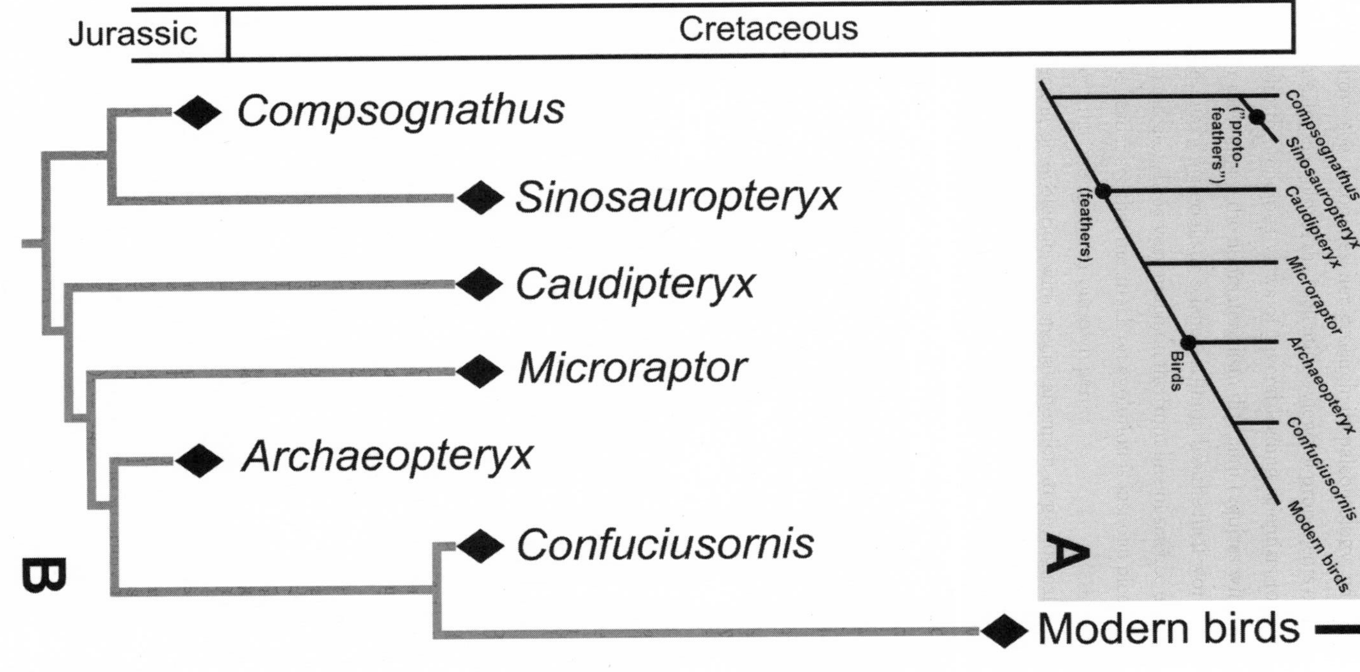

FIGURE 4.1

A. Simplified cladogram depicting the relationships between selected nonavian theropod dinosaurs and birds. B. The same cladogram as shown in A, with branches drawn to different lengths to diagrammatically represent fossil ages. That is, the cladogram shown in A is mapped onto the geologic time scale in B (illustration by author).

Confusion over what a cladogram is and is not lies at the heart of the most common critiques of evolutionary portrayals of bird origins. Recent cladograms depicting hypotheses of bird origins generally show a progression from various small feathered theropod dinosaurs to *Archaeopteryx*, other fossil birds, and finally modern birds (see figure 4.1A). If one reads the cladogram as a genealogy, assuming a progression from ancestors at the base to descendants at the top, there may seem to be an irresolvable flaw in logic because the ages of some of the fossils are not consistent.

Earth scientists divide the Mesozoic era (popularly called the Age of Reptiles) into three consecutive periods: the Triassic (248–206 million years ago), the Jurassic (206–144 million years ago), and the Cretaceous (144–65 million years ago). *Archaeopteryx* is known from rocks of 147 million years ago (Late Jurassic). The Chinese feathered dinosaurs, along with fossil birds more closely related to modern birds than is *Archaeopteryx*, date from roughly 120–125 million years ago (Early Cretaceous). How, therefore, can one claim that the younger feathered dinosaurs are the ancestors of the older *Archaeopteryx*?

The answer is that the feathered dinosaurs are *not* represented as ancestors of *Archaeopteryx;* that is not what the cladogram in figure 4.1A shows. The fact that the Cretaceous dinosaurs appear lower on the cladogram than the Jurassic bird simply means they share fewer bird-like features with modern birds than does *Archaeopteryx*. The feathered dinosaurs are not bird ancestors; they are later descendants of an older common ancestor that they shared with *Archaeopteryx* and other birds. The cladogram shows that, morphologically, the feathered dinosaurs are more distant bird relatives; it claims nothing about their being ancestors of either *Archaeopteryx* or modern birds, because the cladogram doesn't trace direct ancestor-descendant relationships.

Some paleontologists have explained the order of organisms in terms of uncles and cousins (e.g., Padian and Angielczyk 1999), or collateral relatives. In genealogical terminology, a collateral relative is a blood relative, but not an ancestor or a descendant. For example, say that I have a distant cousin, whose great-grandmother was the sister of my own great-grandmother. We are ultimately descendants of a common ancestor (our common great-great-grandparents), but via two different direct lineages. My cousin's relationship to me is that of a collateral relative. She is not part of the same direct ancestor-descendant line as me, but we may share enough features inherited from our distant common ancestor that one can tell we are related and, from our shared characteristics, deduce some of the characteristics of our common ancestor. To bring things back to dinosaurs, consider also the relationship between the protofeathered *Sinosauropteryx* from the Cretaceous, and *Compsognathus*, a Jurassic contemporary of *Archaeopteryx*. Based on currently known theropod fossils, *Sinosauropteryx* and *Compsognathus* are more closely related to each other than either is to other known dinosaurs (see figure 4.1A) (Holtz, Molnar, and Currie 2004), but *Compsognathus* is older (see figure 4.1B). Does this mean that *Compsognathus* is the direct lineal ancestor of *Sinosauropteryx*? Not likely. What it does suggest is that prior to the Late Jurassic, the lineage that led to *Compsognathus* and the lineage that ultimately led to *Sinosauropteryx* diverged from a common ancestor.

The farther apart organisms are on the cladogram, the more distant their relationship. It is true that lower on the tree could popularly be viewed as more "primitive." However, according to cladistics, primitive does not mean *older*; it simply means *fewer derived characters*. Returning to the vertebrate limb example, fish (without four limbs) may be considered more primitive than tetrapods (with four limbs), but this does not mean that all fish have to appear earlier in the fossil record than all tetrapods. New fish species have evolved within the last few millennia in the lakes of the East African Rift Valley. These fish species are younger than our human species *Homo sapiens*, but few would argue they should be placed above humans on the cladogram of vertebrates!

There is no temporal scale on a cladogram; time is not represented. This is not an oversight. Nor is it, as Wells (2000) represents in *Icons of Evolution*, evidence that paleontologists are so determined to present evolutionary series of fossils that they feel free to ignore the order in which organisms appear in the fossil record. Although it is true that the mapping of character distributions onto cladograms is done based solely on morphology without reference to time, this does not mean that the cladogram is incompatible with the fossil record. It is entirely possible to map the cladogram shown in figure 4.1A onto the geologic time scale. One simply extends the branches to different levels, as in figure 4.1B, so that each reaches into the period where the fossils are found. The lengths of the branches approximate the duration of time during which the organisms have been evolving away from their common ancestors, represented by the nodes from which the branches arise. When the cladogram is drawn in this way, it is easier to appreciate that a lower position does not have to equate to an older age. Generally, however, cladograms are drawn with the branches all pruned to the same level, as in figure 4.1A. It is not so much that cladistics ignores time, as that the temporal dimension is not usually depicted.

Returning to the argument that the feathered dinosaurs cannot be bird ancestors because they are younger than *Archaeopteryx*, one can say that this claim is largely irrelevant. No paleontologist is proposing that the Chinese feathered theropods are ancestors of *Archaeopteryx* and later birds. The claim is merely that they are bird relatives, members of lineages that, earlier in the Mesozoic, branched from a common ancestor with the lineage leading to birds.

One nagging question that may remain, however, is why are there no Jurassic feathered dinosaurs? Why do they appear only in the Cretaceous? At least some of the Jurassic dinosaurian relatives of *Archaeopteryx* may very well have been feathered, but the chance of their being recognized as feathered is extremely slim. Feathers and other soft-tissue structures like skin and muscle are extremely difficult to fossilize, because they tend to decay much faster than the skeleton. Only under rare conditions can soft-tissue structures be preserved. *Archaeopteryx* and the feathered dinosaurs of Liaoning are preserved in ancient lake deposits, where quiet water and very slow decay rates allowed preservation of the feathers. It is entirely possible (and in fact very likely) that a feathered dinosaur dying and fossilizing in a different environment would lack all traces of the feathers. Many bird fossils do not preserve feathers either, but no one is suggesting that feathers were not present on the bird while it was alive.

Even following the discoveries of the Chinese feathered theropods, and their acceptance by most of the paleontological community as bird relatives (though not bird ancestors), opponents of evolution have continued to spotlight *Archaeopteryx*. Somewhat ironically, its status within the dinosaur-bird transition has been challenged both on the grounds that it is a bird and on the grounds that it is a dinosaur. Both arguments ring hollow.

ARGUMENT 2

Archaeopteryx *cannot be considered a part of the dinosaur-bird transition because it is a bird.*

Linnaeus first published his now-familiar classification system (Kingdom-Phylum-Class-Order-Family-Genus-Species) in 1735. The system was designed as a hierarchical method of grouping like organisms: the most similar organisms belong to the same species, similar species are grouped into the same genus, similar genera into the same family, and so on. Linnaean classification functioned as a way to group organisms into neat categories based on similarities and differences, as a way for humans to give structure to our understanding of the natural world, and as a way to ease communication among scholars. As new living species were discovered, they were simply placed into the appropriate category with other similar organisms. The system predates Darwin's 1859 publication of *On the Origin of Species by Means of Natural Selection* by more than a century—before Darwinian evolution had been proposed, let alone received widespread acceptance. Linnaeus and his successors were largely operating under a creationist paradigm—there was no widely accepted concept of certain species potentially being ancestors of others, no concept of a branching and ever-evolving tree of life.

As knowledge of the fossil record improved, it became standard practice to try to fit extinct organisms into the established Linnaean hierarchy. Sometimes this worked fairly well, particularly if the organisms resembled modern species. A fossil crocodile, for example, might be easily placed within the vertebrate Class Reptilia, because it resembled modern crocodiles. The discovery of a form distinct from all modern organisms might require the addition of categories to the Linnaean system, but this only required expanding the system, without challenging its entire structure. When Richard Owen named the Dinosauria in 1842, it was clear that nothing like those animals was known to exist in the modern world. They did bear some resemblance to modern reptiles, however, so Dinosauria could be placed within the Linnaean Class Reptilia. As long as fossils fit into the categories erected based on modern organisms, or were distinct from all modern groups, the Linnaean system held up fairly nicely.

The problems began with discoveries of fossils like *Archaeopteryx* that shared characteristics of more than one established group. These forms defied attempts to fit them neatly into the Linnaean system. The difficulty can be illustrated with a simple example. Imagine that you have been given the task of organizing and boxing a collection of colored blocks. You have been given no specific instructions; only that you must group similar blocks together in a logical fashion. The first thing you notice is that the blocks are different colors, so you decide to use color as your primary means of classification. Thinking that this task is absurdly easy, you make rapid progress until you discover a block covered with red and blue stripes. You pause, and wonder where to place this new combination. Should you start a new box for "red and blue" blocks? Should you ignore the blue and assign it to the red box? Does that make more sense than ignoring the red and placing it in the blue box? Can you make a compelling case that red is more important than blue, or vice versa? Even if you do assign it to either the red or the blue box, won't it still be different from its box-mates by virtue of having some of the other color?

The antievolution point of view argues that since *Archaeopteryx* had feathers, presumably flew, and is classified by scientists as the earliest known bird, it must be placed in the bird box. Since it is already a bird, the argument continues, *Archaeopteryx* cannot be transitional between dinosaurs and birds, because something that is already within Class Aves cannot logically be transitional between Class Aves and something outside of Class Aves. What this argument fails to consider is the fact that *Archaeopteryx* is only in Class Aves because humans have decided to place it there. Until the description of the first feathered dinosaurs within the last decade, birds were the only vertebrates known to possess feathers. Therefore, because *Archaeopteryx* had feathers, it was classified as a bird, albeit one whose skeletal structure was otherwise in many ways similar to that of a dinosaur.

Groups in the Linnaean nested hierarchy must be at the same level as or nested within other larger groups. The system does not allow the existence of a category that is transitional between two other categories; therefore, there is essentially no room for transitional forms under the Linnaean system. An organism with an intermediate mosaic of characters is forced to become nontransitional because the classification system *will not allow it* to exist between groups. *Archaeopteryx* is intermediate in its characters, but the system won't allow it to be defined as such. However, the reptilian features of *Archaeopteryx* do not disappear simply because one decides to call it a bird. (Labeling the striped block as red, and placing it in the box with the other red blocks, does not change the fact that it still has some blue.) Setting up a new box for the intermediate does not solve the underlying problem. Class *Archaeopteryx* could not be viewed as intermediate between Class Reptilia and Class Aves, because the boxes would sit next to each other at the same level of the hierarchy.

In one very narrow sense, the antievolutionary argument is technically correct. Under the Linnaean system, Class Reptilia cannot evolve into Class Aves, because they occupy the same level of the hierarchy. However, the fact that one Linnaean class cannot evolve into another does not mean that a dinosaur cannot evolve into a bird. The classes themselves are artificial constructs erected and named by humans trying to organize the natural world into a convenient and logical scheme. Defining Class Reptilia and Class Aves as two separate boxes at the same level in the hierarchy does not mandate that an evolving lineage cannot cross the arbitrary boundary between two classes. When the separate classes Reptilia and Aves were erected, all known reptiles were distinct from all known birds. Linnaean class boundaries are not biological realities; they are lines drawn by humans working with limited information on the diversity of past life.

Viewing the world strictly by the Linnaean system is in many ways incompatible with an evolutionary view of the history of life. This is one of the reasons that modern scientists are moving away from the Linnaean system—there are too many organisms that do not fit nicely into nested boxes. The cladistic methodology preferred by most paleontologists is not concerned with fitting organisms into predetermined boxes. A form with a mosaic of characters will fall out between the organisms or groups of organisms with which it shares its characters. It does not have to be forced into one of those groups, and there is no need to create a new group to accommodate it. It is perfectly acceptable for it to exist as what it is—a form with a mosaic of characters.

A further important point is that the Linnaean and cladistic systems should not be mixed, because they are fundamentally different ways of conceptualizing organisms and their relationships. One cannot overlay Linnaean class boundaries on the cladogram of bird origins and use this to prove that the cladogram is wrong. It has been argued that the feathered dinosaurs cannot be closely related to birds because the former are members of Class Reptilia and the latter are from Class Aves, and animals in different classes cannot be part of the same lineage (Wells 2002). This argument has no logical basis, because the Linnaean classes themselves do not recognize lineages, and cladistics doesn't recognize the Linnaean class boundaries. One cannot argue that one system's hypothesis is false based on the criteria of a completely different system.

One might assume that a better understanding of cladistics, and the ways in which it is distinct from the Linnaean classification system, would further the acceptance of transitional forms on the dinosaur-bird line. This, however, has not necessarily been the case.

ARGUMENT 3

Archaeopteryx *cannot be considered a part of the dinosaur-bird transition because it is a dinosaur.*

The cladistic system isn't concerned with shoehorning organisms into boxes, but that is not to say it doesn't recognize named groups. According to cladistics, a named group includes within it all the descendants of that group's common ancestor. Following the hypothesis that birds evolved from dinosaurs, birds technically are dinosaurs. Although it may seem a little extreme to think of a pigeon as a dinosaur, this simply reflects the naming conventions of cladistics. The pigeon is still a bird, because the group *bird* generally includes *Archaeopteryx*, living birds, and all the descendants (living and extinct) of their common ancestor (Padian 2004). But the pigeon is also a dinosaur, because the group *dinosaur* includes all the descendants of the common ancestor of all dinosaurs, and this group would have to include birds if birds are descended from dinosaurs. The group *birds* still exists, but birds represent a subset of *dinosaurs*. Following this logic, if birds are dinosaurs, then *Archaeopteryx*, even though it is a bird, is also a dinosaur.

In *Icons of Evolution*, Wells (2000) argues that the cladistic view of birds as dinosaurs defines *Archaeopteryx* out of a transitional position, because it becomes just another dinosaur. To say that calling *Archaeopteryx* a dinosaur makes it unqualified to be a transitional form makes no more sense than saying it cannot be a transitional form if it is classified as a bird. *Archaeopteryx* has a mosaic of dinosaurian and avian features. Classifying it as a dinosaur does not wipe away its avian characteristics, no more than classifying it as a bird wipes away its many dinosaurian features.

As an aside, *Icons of Evolution* also treats the entire cladistic convention of calling birds dinosaurs as absurd, saying that one might as well claim that humans are fish. This criticism is somewhat ironic because, according to the naming conventions of cladistics, technically humans *are* fish. This is not to say that humans have fins and gills and live in the ocean. If tetrapods are descended from fish, then tetrapods are an extremely derived subset of fish. If humans are mammals, and mammals are a group of tetrapods, then humans are technically fish, because the group *fish* would include all the descendants of the common ancestor of all fish, and this would include all tetrapods. This again is an artifact of the way cladistic groups are defined, and really becomes ridiculous only if one insists on trying to overlay the Linnaean system of five separate and equally ranked vertebrate classes (fish, amphibians, reptiles, birds, mammals) onto the cladistic system. Humans are still mammals, but the group *mammal* no longer exists at the same level as *fish*, since ultimately mammals and all other tetrapods can be traced to an ancestor within *fish*. The organisms and their characteristics have not fundamentally changed; we are simply classifying them in a different way.

Regardless of whether one chooses to call *Archaeopteryx* a bird, a dinosaur, or both a bird and a dinosaur, its anatomy remains unchanged. Most antievolutionary arguments will readily concede that *Archaeopteryx* possesses a mosaic of features.

They will not, however, grant it transitional status. Why? According to their definitions, a mosaic is not transitional.

ARGUMENT 4

Neither Archaeopteryx *nor the Chinese feathered dinosaurs are relevant because a transitional form should be intermediate between two groups in all its features, rather than simply having a mosaic of both groups' features.*

When discussing transitional forms in the dinosaur-bird line, paleontologists and opponents of evolution are considering the same set of fossils. How can the same organisms support both sides of the argument? A large part of the difference hinges on a semantic argument over the meaning of *transitional* (Cracraft 1983; Eldredge 2000). From the paleontologist's point of view, a form that is transitional between two groups would be expected to display some of the features that characterize one group, and some of the features that characterize the other. Opponents of evolution, on the other hand, hold that the condition of a feature in a transitional form must lie partway between the condition of that specific feature in the first group and its condition in the second group. (It is not enough for a transitional block to have red and blue stripes; it must be purple.) *Archaeopteryx* can be a bird with dinosaur-like features, but having some avian and some dinosaurian features is not enough to grant it transitional status. For *Archaeopteryx* to be considered transitional under this definition, it must have features that are not dinosaurian and not avian, but something between the two.

Why *must* a transitional form be a blend; why can it *not* be a mosaic? Why recognize only one narrow definition of transitional? Consider that species in one of several lineages of *bird-like dinosaurs* over time accumulate more and more avian features, so that eventually some species are recognizable as *dinosaur-like birds*, and finally simply as *birds*. Why is this a more improbable transition than the expectation of finding one species that is intermediate in all respects?

Perhaps one does not even need to be concerned with the semantics of mosaic versus blended features. The simplest way to appreciate the existence of transitional forms may be to consider the fossils that have had their dinosaurian or avian identities questioned and/or changed. A specimen of *Archaeopteryx* without clear feather impressions was misidentified as the small theropod *Compsognathus* (Wellnhofer 1974). One group of researchers argues that some of the feathered dinosaurs are actually birds (Martin and Czerkas 2000; Feduccia 2002, 2005). Members of one enigmatic group of theropods (the alvarezsaurids) were originally described as birds (Perle et al. 1993), only later to be moved back to nonavian status (Sereno 2001), and their actual position is still debated. Opponents of evolution have used these examples as evidence against a close relationship between dinosaurs and birds (e.g., Sarfati 1999). But would it not make more sense to argue that such confusing anatomy is good evidence *for* a close relationship?

The point is not the specific arguments about what should be called a dinosaur and what should be called a bird. The point is rather that, when it comes to drawing a line between dinosaur and bird, there are several species that can fall on either side, depending on how one interprets the anatomy. If paleontologists are having this much trouble drawing a line between dinosaur and bird, doesn't that itself argue for the lack of a neat division between the two groups?

SUMMARY

Discussion of the dinosaur-bird transition has figured prominently in the debate over vertebrate evolution and the existence of transitional forms in the fossil record. Opponents of evolution have advanced four principal arguments contesting the status of *Archaeopteryx* and the feathered dinosaurs as key players in the dinosaur-bird transition. Three of those arguments largely stem from misinterpretation of the cladistic methodology used by most paleontologists to determine relationships between organisms. No paleontologist has claimed that the Cretaceous feathered dinosaurs are the ancestors of the Jurassic bird *Archaeopteryx*. Cladograms are not family trees, and they are not meant to represent a chronological progression of species through time, nor are they attempts to misrepresent or ignore the order in which species appear in the fossil record; they are simply maps of distributions of shared derived characters. Whether *Archaeopteryx* is called a bird or a dinosaur does not erase the fact that it possesses a mosaic of dinosaurian and avian characters. Finally, there is no compelling reason to consider a mosaic any less transitional than a form with blended intermediate features. One could argue that such a narrow definition of transitional becomes less and less tenable with the continuing discovery of additional species that further blur the boundaries between what we call dinosaur and what we call bird. The fossils speak for themselves. That they do not always clearly identify themselves with one group or the other is persuasive evidence for lineages in transition through time.

It is interesting to note that after devoting an entire chapter in *Icons of Evolution* to the discussion of *Archaeopteryx* and bird origins, Wells makes no mention of either in his more recent book *The Politically Incorrect Guide to Darwinism and*

Intelligent Design (2006) and confines his arguments against intermediate forms to a discussion of whale origins. On the one hand, Wells may simply believe his previous arguments need no repetition. Up to this point, however, *Archaeopteryx* has been a nearly ubiquitous staple in arguments against vertebrate evolution. Could it be that, as more and more fossils of feathered dinosaurs and early birds are discovered and described, it is becoming increasingly difficult to contradict what the fossils are saying?

REFERENCES

Behe, M. J. 1996. *Darwin's Black Box: The Biochemical Challenge to Evolution*. Free Press, New York.

Cracraft, J. 1983. Systematics, Comparative Biology, and the Case against Creationism. Pp. 163–92 in L. R. Godfrey, ed., *Scientists Confront Creationism*. Norton, New York.

Darwin, C. 1859. *On the Origin of Species by Means of Natural Selection*. John Murry, London.

Eldredge, N. 2000. *The Triumph of Evolution . . . and the Failure of Creationism*. Henry Holt, New York.

Feduccia, A. 2002. Birds are Dinosaurs: Simple Answer to a Complex Problem. *Auk* 119(4): 1187–1201.

———. 2005. Do Feathered Dinosaurs Exist? Testing the Hypothesis on Neontological and Paleontological Evidence. *Journal of Morphology* 266:125–66.

Gish, D. T. 1973. *Evolution: The Fossils Say No!* Creation-Life, San Diego, CA.

Gishlick, A. D. 2004. Evolutionary Paths to Irreducible Systems: The Avian Flight Apparatus. Pp. 58–71 in M. Young and T. Edis, eds., *Why Intelligent Design Fails: A Scientific Critique of the New Creationism*. Rutgers University Press, New Brunswick, NJ.

Holtz, T. R., Jr., R. E. Molnar, and P. J. Currie. 2004. Basal Tetanurae. Pp. 71–110 in D. B. Weishampel, P. Dodson, and H. Osmólska, eds., *The Dinosauria*, 2nd ed. University of California Press, Berkeley.

Ji, Q., P. J. Currie, M. A. Norell, and S.-A. Ji. 1998. Two Feathered Dinosaurs from Northeastern China. *Nature* 393:753–61.

Ji Q. and S. Ji. 1996. On the Discovery of the Earliest Fossil Bird in China (*Sinosauropteryx* Gen. Nov.) and the Origin of Birds [in Chinese]. *Chinese Geol.* 233:30–33.

Koons, R. C. 2004. The Check Is in the Mail: Why Darwinism Fails to Inspire Confidence. Pp. 3–22 in W. A. Dembski, ed., *Uncommon Dissent: Intellectuals Who Find Darwinism Unconvincing*. ISI Books, Wilmington, DE.

Martin, L. D., and S. A. Czerkas. 2000. The Fossil Record of Feather Evolution in the Mesozoic. *Am. Zool.* 40(4): 687–94.

Meyer, H. von. 1861. *Archaeopteryx lithographica* (Vogel-Feder) und *Pterodactylus* von Solnhofen. *N. Jb. Min. Geol. Paläeontol.* 678–79.

Morris, H. R. 1974. *Scientific Creationism*. Creation-Life, San Diego, CA.

Owen, R. 1842. Report on British Fossil Reptiles, part 2. *Rep. Br. Assoc. Adv. Sci.* 11:60–200.

Padian, K. 2004. Basal Avialae. Pp. 210–31 in D. B. Weishampel, P. Dodson, and H. Osmólska, eds., *The Dinosauria*, 2nd ed. University of California Press, Berkeley.

Padian, K., and K. D. Angielczyk. 1999. Are There Transitional Forms in the Fossil Record? Pp. 47–82 in P. H. Kelley, J. R. Bryan, and T. A. Hansen, eds., *The Evolution-Creationism Controversy II: Perspectives on Science, Religion and Geological Education*. Paleontological Society, Fayetteville, AR.

Perle, A., M. A. Norell, L. M. Chiappe, and J. M. Clark. 1993. Flightless Bird from the Cretaceous of Mongolia. *Nature* 362:623–26.

Robertson, M. 2004. Chinese Feathered Dinosaurs: Where Are the Skeptics? *Answers in Genesis*, July 13. www.answersingenesis.org/home/area/chinesedinosaurs/featheredDinos.asp.

Sarfati, J. 1999. *Refuting Evolution: A Handbook for Students, Parents, and Teachers Countering the Latest Arguments for Evolution*. Master Books, Green Forest, AR.

Sereno, P. 2001. Alvarezsaurids: Birds or Ornithomimosaurs? Pp. 69–98 in J. Gauthier and L. F. Gall, eds., *New Perspectives on the Origin and Evolution of Birds*. Yale University Press, New Haven, CT.

Wellnhofer, P. 1974. Das fünfte Skelettexemplar von *Archaeopteryx*. *Palaeontographica Abt. A.* 147:169–216.

Wells, J. 2000. *Icons of Evolution: Science or Myth? Why Much of What We Teach about Evolution Is Wrong*. Regnery, Washington, DC.

———. 2002. Inherit the Spin: Darwinists Answer "Ten Questions" with Evasions and Falsehoods. *Discovery Institute*, January 15. www.discovery.org/scripts/viewDB/index.php?command=view&id=1106.

———. 2006. *The Politically Incorrect Guide to Darwinism and Intelligent Design*. Regnery, Washington, DC.

Xu, X., Z. H. Zhou, X. L. Wang, X. W. Kuang, F. C. Zhang, and X. G. Du. 2003. Four-Winged Dinosaurs from China. *Nature* 421:335–39.

PART TWO · EDUCATION, POLITICS, AND PHILOSOPHY

FIVE ·

Pangloss, Paley, and the Privileged Planet

Parrying the Wedge Strategy in Earth Science Education

MARK TERRY

When our '85 Vanagon broke down not far from our house during the summer of 2006, I didn't expect it to lead to a discussion about intelligent design (ID). Believing the problem to be the fuel pump and hoping for a quick fix, I called for a tow to the only nearby VW shop open on Saturdays. Not five minutes after picking me up, the tow-truck driver brought up the topic of intelligent design. He had discovered I was a science teacher, and expressed the hope that I was *not* having to deal with it in my classroom. We had a great conversation and parted friends. He had his own backyard ID problem, and I ended up sympathizing with him. An avid amateur astronomer, he was trying to deal cordially with his new neighbor, who had set about trying to convince him that there is scientific evidence that ID permeates the universe. As it turned out, my fuel pump theory wasn't supported by the evidence, a rare part was ordered, and I might as well have delayed the tow until the next week—but I'm glad I didn't miss that conversation.

Intelligent design was in the air that summer, thanks to a dynamite combination of the marketing smarts and experience of the Discovery Institute, the fundamentalist leanings of about a third of the American public, and an even more broadly shared scientific illiteracy. Astoundingly, ID continues to loom over discussions by public school science teachers, school board members, electoral candidates, and neighbors despite having virtually no presence in scientific research journals. Much has already been written about how ID is antithetical to good science education (Forrest and Gross 2004; Terry and Linneman 2003). Across the country most science teachers, university professors, and researchers largely find its popularity bewildering and annoying, and wish it would all go away (Pennock 2001; Terry 2004). They might as well pine for the demise of fast food.

The boundless energy of the ID campaign inspires awe. Money has flowed to it steadily since the mid-1990s, and the very fundraising document that stimulated this can help us understand why. The Wedge Document is a carefully crafted case statement designed to rally big donors around a well-focused cause. First intended for internal eyes only, then later broadcast on the Web without authorization, the Wedge Document was ultimately claimed by the Discovery Institute as its own (Downey 2006). It calls for challenging public school teaching of evolution and instead infusing the curriculum with ID as a means to a larger end: opening statewide curricular definitions of science to include the supernatural. In turn, this new definition of science is intended to help cement Christian religious principles, as understood by the authors of the Wedge Document, at the center of American civic, legal, scientific, and religious life (Center for the Renewal of Science and Culture). Directing this Wedge strategy, as it is named in the document, is the Center for Science and Culture (CSC, originally named the Center for *the Renewal of* Science and Culture) of the Discovery Institute in Seattle, Washington. The institute is otherwise known around the Pacific Northwest as a civic affairs think tank with Christian conservative leanings that makes occasional contributions to discussions of public transit, regional government, and other issues (Killen 2003; Scigliano 2006).

The Discovery Institute's CSC has managed to unfurl the huge canvas of a so-called Big Tent for creationists of all stripes, the Christian Right, and other political conservatives (Nelson 2002; Pennock 2000). From this tent emanates a much broader nationwide attack on the teaching of evolution than at any time in the past. Focus on the Family, the conservative Christian organization led by radio personality and writer James Dobson, produces the institute's "science" videos with slick technical production. InterVarsity, Regnery, and other Christian presses publish the bulk of its books (though lately the institute has begun publishing on its own), and Reasons to Believe, a creationist organization that claims astronomy proves the existence of God, hawks the institute's wares at science teacher conventions. Feeling the public relations force of this attack, public school science teachers, administrators, and school board members are surrounded by a confusion of ill-defined terms and by accusations that they are just too ignorant to understand the new cutting edge of ID in biology, Earth science, and indeed all of the natural sciences (Terry 2005b).

With the congregation in the Big Tent lustily singing the praises of ID and preaching against the evils of Darwinism, trying to ignore ID fails at the outset and only serves to permit the advance of the Wedge strategy. Attempting to keep ID out of classrooms, as with any prohibition, only heightens interest. The Discovery Institute gleefully seizes on any efforts to limit access to ID as fairness and free speech infringements, which suits the overall goals of the Wedge strategy even better than the actual inclusion of ID in the curriculum. Righteous indignation about "viewpoint discrimination" can be stirred up without any science. If scientists can be portrayed as censoring the "new idea," as in the 2008 ID movie *Expelled. No Intelligence Allowed*, all the better (Miller and Stein 2008). America loves the underdog and the upstart (Brauer, Forrest, and Grey 2005).

I've been teaching high school biology for over thirty years, and my classes always have a strong evolutionary content. When the Discovery Institute opened its headquarters just a few blocks from my school, I had no idea that natural theology, which we'd always included for historical purposes in our curriculum, was about to be championed again as a science. Watching the success of the institute's public relations campaign has convinced me there are major forces at work. I don't think any of us should fool ourselves that this educational-political-religious problem will be solved any time soon. Meanwhile, science teaching and the public understanding of science are in need of serious help! To that end, in this essay I will recommend a historical, interdisciplinary approach that can provide a useful perspective on ID's role in the development of modern science (Terry 2005a, 2005c, 2005d). The Earth science content of a major contribution to the modern ID bookshelf, *The Privileged Planet* (Gonzalez and Richards 2004), will be evaluated in light of this historical context. The Wedge strategy is most effective when its message slips into mainstream science literature. There's a prime example of this in a recent journal article about Earth science education (Ross 2005), which if studied carefully can become an effective resource for teachers about the Wedge strategy itself.

EARTHQUAKES BY DESIGN

Pre-Enlightenment approaches to the natural world can be useful to show modern science in sharp relief. A brief step back in time invites such questions as "Plate tectonics, instead of what?" "Mendel's genetics, instead of what?" "Natural selection, instead of what?" This is particularly important today, since the great sorting of ideas that took place during the Enlightenment, roughly from the late seventeenth to the early nineteenth century, resulted in the abandonment of ID, referred to at the time as "natural theology." The "argument from design" is not a novel, cutting-edge scientific insight, as the Discovery Institute would have us believe, but an approach with an honorable and interesting place in the history of ideas—an idea that inspired many, including Darwin, to study the natural world in greater detail (Brooke 1991).

Detailed studies ultimately led away from natural theology to a modern science that deals only with natural causes and does not comment on the existence or characteristics of powers or designs that might lie outside nature. Selected quotations or illustrations from period literature reflect beliefs that changed over time. For example, an effective way to highlight the Enlightenment transition away from the natural theology version of ID in the Earth sciences is to focus on the consideration of earthquakes, from Voltaire's eighteenth-century *Candide* to their treatment by Lyell and Darwin.

Before the Enlightenment, scientists and nonscientists commonly believed that earthquakes were designed to teach us about human behavior, good and evil, rather than the Earth's structure. Both the living and nonliving worlds were assumed to have moral content, and to observe them correctly was to receive instruction or admonition from God (White 1954). Voltaire pointedly placed the 1755 Lisbon earthquake in *Candide* to highlight and mock this kind of thinking. Voltaire was in the process of abandoning the natural theology view of the Earth in part *because* of the earthquake (Durant and Durant 1965, pp. 720–26). Happily, *Candide* is still widely read in high schools and introduces many students to the European Enlightenment. This biting satire can also provide an introduction to the shift in thinking necessary for the development of modern earthquake studies.

As the narrative passes through one calamity after another, Candide and his tutor, Dr. Pangloss, eventually go from a shipwreck to apparent salvation in Lisbon, only to find themselves almost immediately in one of the most destructive earthquakes in recorded history. They survive, but the seemingly senseless destruction of innocent lives gives yet another opportunity for Pangloss to proclaim that this is the best of all possible worlds: "For all this is for the very best. For if there is a volcano in Lisbon, it could not be anywhere else. For it is impossible that things should not be where they are. For all is well" (Voltaire 1759, p. 26). Subsequently the local university arranges the burning of a few outcast individuals, "an infallible secret for keeping the Earth from quaking" (p. 27).

It was precisely the moralistic reasoning of the design arguments of his day regarding the Lisbon earthquake that led Voltaire to portray the natural theology position with such sarcasm. Voltaire found repugnant the presumption of moral, divine causation for such an event; he therefore constructed the character Pangloss to satirize what he perceived to be this weak-minded approach to explaining natural phenomena. Voltaire saw the *absence* of design in the earthquake that struck Lisbon on a Sunday morning when thousands of devout worshippers were crushed in their great stone churches. In a poem published the year after the earthquake, Voltaire expressed outrage at the proposed connection between human behavior and Earth processes:

Oh, miserable mortals, grieving earth!
Oh, frightful gathering of all mankind!
Eternal host of useless sufferings!
Ye silly sages who cry, "All is well,"
Come, contemplate these ruins horrible,
This wreck, these shreds and ashes of your race;
Women and children heaped in common death,
These scattered members under broken shafts;
A hundred thousand luckless by the earth
Devoured, who, bleeding, torn, and still alive,
Buried beneath their roofs, end without help
Their lamentable days in torment vile!
To their expiring and half-formed cries,
The smoking cinders of this ghoulish scene,
Say you, "This follows from eternal laws
Binding the choice of God both free and good"?
Will you, before this mass of victims, say,
"God is revenged, their death repays their crimes"?

(VOLTAIRE 1756, pp. 721–22)

Of course, the design argument persisted beyond Voltaire. In fact, its most notable expression was soon to be published as William Paley's clear, readable, and immensely popular *Natural Theology* (1802). Paley proposes that the Earth was well made by God for human purposes, and that both the existence and character of God could be known through God's works. Paley devotes most of his argument to what he considers the best of all evidence, namely the wondrous anatomy and physiology of living things. He does not discuss earthquakes, but he does make claims about some of the physical constituents of our planet as well as its position in the solar system. Water, for example, he describes as perfect. Explorations of its molecular structure, which began during Paley's lifetime, afforded him the opportunity to comment on the utility of such investigations:

> But then it is for our comfort to find, that a knowledge of the constitution of the elements is not necessary for us. For instance, as Addison has well observed, "we know water sufficiently, when we know how to boil, how to freeze, how to evaporate, how to make it fresh, how to make it run and spout out in what quantity and direction we please, without knowing what water is." The observation of this excellent writer has more propriety in it now, than it had at the time it was made: for the constitution and the constituent parts of water, appear in some measure to have been lately discovered; yet it does not, I think, appear, that we can make any better or greater use of water since the discovery, than we did before it. (Paley 1802, pp. 207–8)

This passage epitomizes how the design argument discourages deep inquiry.

The greatest contribution to the design argument in geology during the nineteenth century was William Buckland's 1836 volume in the series The Bridgewater Treatises. Buckland notes the perfection visible in the sedimentary deposits resulting from Earth processes. He maintains that such deposits, by design, provide all manner of benefits for human beings. For example, Buckland comments on the placement of coal seams for the anticipated use of the British:

> It is impossible to contemplate a disposition of things, so well adapted to afford the materials essential to supply the first wants, and to keep alive the industry of the Inhabitants of our earth; and entirely to attribute such a disposition to the blind operation of Fortuitous causes. . . . We may surely therefore feel ourselves authorized to view, in the Geological arrangements above described, a system of wise and benevolent Contrivances, prospectively subsidiary to the wants and comforts of the future inhabitants of the globe; and extending onwards, from its first Formation, through the subsequent Revolutions and Convulsions that have affected the surface of our Planet. (1836, p. 547)

In contrast, Charles Lyell and his geological disciple Charles Darwin represent the Enlightenment transition away from natural theology; there is no suggestion in their writings of design or supernatural causation of earthquakes or other geological events. Instead, in the first volume of his *Principles of Geology* (1830) Lyell recounts an exhaustive list of earthquake reports, including the Lisbon quake, much like a modern physician might prepare a long list of symptoms and outbreaks to try to understand an infectious disease (Lyell 1830, pp. 399–479). Though Lyell was not the first to look for natural causes in geology, his writings mark the maturation of the approach: "Geology is the science that investigates the successive changes that have taken place in the organic and inorganic kingdoms of nature; it inquires into the causes of these changes and the influence which they have exerted in modifying the surface and external structure of our planet. By these researches into the state of the earth and its inhabitants at former periods, we acquire a more perfect knowledge of its present condition, and more comprehensive views concerning the laws now governing its animate and inanimate productions" (Lyell 1830, p. 1).

As for the young Darwin, most of his interpretations of nature prior to his life-altering voyage on the H.M.S. *Beagle* were aligned with Paley's *Natural Theology*. On the outgoing leg of the voyage, however, Darwin read Lyell's first volume and absorbed the treatment of earthquakes and the argument for uniformitarian interpretation in geology (Browne 1995, pp. 93, 186–90). He then experienced a large earthquake firsthand on the coast of Chile. Darwin's detailed description of the earthquake makes clear that he intends to follow in the footsteps of Lyell, looking for unvarnished data first, then for evidence of geologic laws. Nowhere does Darwin question the justice or purpose of the earthquake that affected the Chileans, nor does he judge Chileans and their society. He waxes philosophical once simply to enhance his description of the experience: "A bad earthquake at once destroys our oldest associations: the earth, the very emblem of solidity, has moved beneath our feet like a thin crust over a fluid;—one second of time has created in the mind a strange idea of insecurity, which hours of reflection would not have produced" (Darwin 1836, pp. 260–69). Darwin then describes every bit of physical evidence of earthquake damage he can without even a passing reference to any deity or to design.

As can be seen by reading Darwin and Lyell, science during the Enlightenment stepped away from the search for supernatural causes and design. Scientists could not refute design for they could not put restrictions on what a force outside nature could do. But seeing what Paley and Buckland either overlooked or chose not to mention, they pursued other questions, such as: what mechanisms operate in the natural world? By asking such questions, Darwin, Lyell, and others moved beyond

Paley and Buckland methodologically. Design was abandoned as a method of explanation and analysis for science, and scientists have steadily searched for natural causes ever since. This shift in approach can quickly be apprehended by students through reading and discussion of selected brief excerpts from Paley, Buckland, Lyell, and Darwin.

REDISCOVERING DESIGN

Whereas modern science searches for mechanisms and natural causes for natural phenomena, in *The Privileged Planet* (2004) astronomer Guillermo Gonzalez and philosopher Jay W. Richards, both senior fellows of the Discovery Institute's CSC, search for a purpose, as did their antecedents, William Paley and William Buckland. In a new twist, they assert that ID is demonstrated by our scientific successes themselves. Their argument goes like this: if the universe and our planet were not the way we have discovered them to be, we wouldn't be able to do the science that has discovered those characteristics of the universe and our planet. They call this "discoverability" (Gonzalez and Richards 2004, pp. 210–12). Discovery senior fellow Jonathan Witt coauthored *The Privileged Planet* video released to accompany the book, and stated the discoverability principle: "The same narrow circumstances that allow us to exist also provide us with the best overall setting for making scientific discoveries" (Allen and Witt 2004, ch. 8). For example, Gonzalez and Richards attribute the following design aspects to earthquakes. Because they are well distributed on diverse continents on opposite sides of the globe today, earthquakes allow us to understand the structure of the Earth. Gonzalez and Richards state that we couldn't have understood the Earth as well during the time of Pangaea, when all of today's continents were clustered together in one supercontinent. Earthquakes at lithospheric plate margins would have been less frequent owing to the dearth of margins, and poorly distributed owing to the lopsided placement of the single continent on the globe. But humans weren't around then either. Now we're here; earthquakes are frequent and well distributed, and as a result we are permitted to discover Earth's structure (Gonzalez and Richards 2004, pp. 45–48, 60). However, the fact that continents were also spread across the globe in pre-Pangaean time, when earthquakes would have been frequent and well distributed but with no humans around to make seismic measurements and study the Earth, at least raises the question of what "discoverability" could possibly mean in the absence of discoverers.

Gonzalez and Richards state that earthquakes were designed for discovery, and in an optimistic twist that may be a nod to the human suffering that has come with this "scientific convenience," the authors suggest that to *reduce* human suffering, populations simply must retreat from tectonically active coastal-zone margins (Gonzalez and Richards 2004, p. 362, n. 18). Earthquakes are designed, they say, and the designer has supernatural powers sufficient to arrange continents, cause them to move about the surface of the Earth, and time the appearance of humans on the scene. It is an awesome prospect, though whether it is better to be crushed in an earthquake that's part of the discoverability plan or in one simply meted out as divine punishment is difficult to say.

A striking revelation in *The Privileged Planet* and its companion video is that near-total solar eclipses are made possible because of the proportions, distances, and relative movements of the sun, moon, and Earth. Total eclipses pack emotional power, much as the monolith alignment sequences in Stanley Kubrick's 1968 science fiction classic *2001: A Space Odyssey* (Kubrick and Clarke) do, or as does the eclipse prediction made by Twain's Connecticut Yankee, which cemented the power of the nineteenth-century common man over his medieval antecedents (Twain 1889, pp. 33–41). In *The Privileged Planet*, Gonzalez and Richards claim that such eclipses must have been designed by arranging the right-sized objects of the sun, moon, and Earth, at the correct relative distances in a straight line. Humans are around to observe them and discover solar elemental composition through spectral analysis of the resulting solar halo (2004, pp. 310–11). The objection that the relative positions of the bodies are steadily changing, as the moon drifts farther from Earth, is easily rebuffed by the assertion that humans are here now because the timing has been arranged by the designer.

The search for purpose takes place in realms other than the scientific. Surely planets or moons with shifting continents and "well-placed" active earthquake zones may exist whether or not seismologists exist to observe them. The presence of human beings on planet Earth at this exact time proves nothing either for or against purpose or design, no matter how inspiring the experience of a total solar eclipse may be. "Discoverability," as evidence for design, at least comes packaged by a religious publishing house (Regnery) and video producer (Illustra, Focus on the Family), and its authors display their Discovery Institute affiliations. Despite claims of a scientific breakthrough, the circularity and religious foundations of the argument are fairly obvious. It's not so difficult to parry such a direct thrust of the Wedge, and it would be easy to use short clips from the *Privileged Planet* video to introduce the ID strategy to a class.

THE GEOLOGISTS' PICK VERSUS THE WEDGE

In contrast to *The Privileged Planet*, an article published in the *Journal of Geoscience Education* in 2005 exemplifies a subtler Wedge strategy tactic. The author, Marcus Ross, received front-page coverage in the *New York Times* for earning his Ph.D. from the Department of Geosciences at the University of Rhode Island while being a young-Earth creationist. This feat made an intriguing story, since Ross's research was on seagoing reptiles, the mosasaurs, that paleontologists report as having gone extinct some 65 million years ago. The headline read, "Believing Scripture but Playing by Science's Rules" (Dean 2007).

But Ross's *Journal of Geoscience Education* article, "Who Believes What? Clearing Up the Confusion between Intelligent Design and Young-Earth Creationism," does not advocate playing by the rules of science (Ross 2005). In fact, it advocates a view of science straight out of the pre-Enlightenment period, infusing sectarian religious ideas into science, one of the goals of the Wedge strategy. The article, based in part on a talk given by Ross at the November 2003 Geological Society of America meeting in Seattle (Ross 2003), invites its readers to improve geoscience teaching about evolution by introducing students to a nested hierarchy of beliefs about purpose in the universe. It also suggests that character-state principles, usually used to distinguish anatomical specializations among fossils, can be applied to belief systems in order to introduce the difficult concept of cladistic analysis in a paleontology unit or class (Ross 2005). On the surface, this all seems reasonable, useful, and innovative, and undoubtedly the article was accepted for publication on these grounds.

But Ross was once a fellow of the Discovery Institute's CSC (Forrest and Gross 2004, p. 59). His article was subsequently redrafted and coauthored by young-Earth creationist and CSC fellow Paul Nelson (Ross and Nelson 2006), and on close inspection is revealed to have significant Wedge strategy content. Ross proposes nine "positions," his summaries of beliefs regarding the presence of purpose or design in the natural world, from "Materialist Evolutionist" to "Young-Earth Creation." As Canadian media critic Marshall McLuhan famously instructed, "the medium is the message" (1964). While seeming to focus on improving education techniques for Earth science, Ross has succeeded in placing the following capitalized terms in the text of a peer-reviewed science teachers' journal article: "Being," "Deistic Being," "Theistic Being," and "God." The article even raises a discussion of the true intended meaning of John in the book of Revelation of the Christian New Testament. These terms and questions would have made sense to Buckland in 1836, but geology shifted its course and followed Lyell, who would have argued as early as 1830 that such questions have no place in a scientific investigation.

Ross also promotes a destructive false dichotomy of the sort that needlessly drives the religious faithful away from science: "To justify a science/nonscience demarcation, it must be shown that the Bible and science are mutually exclusive. It follows that if the Bible is non-science, then the Bible cannot now, nor ever have, provided any framework for scientific investigation. Neither can it aid in generating any testable hypothesis. If strict demarcation is true, then a scientist cannot use the Bible to gain meaningful insight while in the pursuit of scientific knowledge" (2005, p. 320). Yet the literature of science is strewn with examples of scientific insights coming from all manner of sources, religious and nonreligious. Science *must* be demarcated from nonscience, but scientists themselves may be inspired by almost anything: a cloud in the sky, a musical phrase, a punch in the nose, a religious text (Bronowski 1973; Brooke 1991).

Ross then claims that a signal advantage of his analysis is that "it is not intended to distinguish which of the positions classified can be referred to as 'scientific' positions" (2005, p. 321). This assertion removes the necessity of acknowledging that intelligent design is not science, because whether or not it is science is declared *not to be an issue*. That's an extraordinary claim to be made in a science teachers' journal, but it serves the Wedge strategy goal of establishing "a science consonant with Christian and theistic convictions" (Center for the Renewal of Science and Culture, p. 6).

Ross wrote a near-perfect Wedge article: one that proposes that there is no necessary clear distinction between natural and supernatural explanations of the natural world. Publication of the article resulted in enough controversy within the National Association of Geoscience Teachers that careful consideration was given to its possible retraction. Wisely, the association decided not to retract it (Feiss 2007), because a whiff of ideological censorship could always be conjured up around such an action, similar to what happened when CSC director Stephen Meyer's article on intelligent design was first published, then retracted from the *Proceedings of the Biological Association of Washington* (Numbers 2006, pp. 390–91).

Future reviewers may watch more closely for Wedge strategy content, now that Dr. Ross is an assistant professor in the Biology/Chemistry Department at Jerry Falwell's Liberty University in Lynchburg, Virginia. After all, the university publishes a doctrinal statement that asserts that "the universe was created in six historical days" and "human beings were directly created, not evolved" (Liberty University 2007).

SCIENCE EDUCATION IN THE TIME OF THE WEDGE

Teachers, parents, school board members, and interested citizens need to be aware of the workings of the Wedge strategy throughout the sciences, including Earth science, from junior high schools to colleges to museums. Ideas have histories, and the historical roots of ID in natural theology can help demystify and identify the modern movement. Interdisciplinary collaborations employing the history of science can help enlighten students and the general public in a way that avoids discrediting ID as a personal philosophy or belief, but shows unambiguously that science gave up this approach for a much more productive one centuries ago.

All of us need to be ready to "bell the cat," as were a group of parents and the science teachers of the Dover School District in 2005 (Humes 2007, pp. 214–16). If an ID production, such as *The Privileged Planet*, is promoted or displayed as a work of science in a school or library, parents, teachers, and citizens of all sorts need to call for discussions about whether or not this is a scientific or a religious work. A look to the positions and statements of the publishers and authors in such a case should carry the day, much as the close review of the ID supplemental text *Of Pandas and People* was pivotal in the Dover School District trial (Humes 2007, pp. 284–87). An article such as Ross's, reviewed earlier, can provide a great stimulus for discussion of the Wedge strategy itself. The subtlety of its approach is mirrored in that of the design community's proposed changes to science education standards in Kansas, which would have promoted the exploration of supernatural causes in public school science classes (Terry 2007, pp. 44–46).

Of course it would be wrong to suppress ID as a religious idea. But not suppressing a religious idea and labeling it science are two different things. Science does not assert that it is wrong to believe that the universe or any aspect of it was designed. Science does not evaluate claims about the supernatural. What *is* wrong is to allow supernatural explanations in the natural sciences, or in a gesture of "fairness" to require science educators to present supernatural claims alongside scientific observations.

Paley was not wrong to believe in his version of intelligent design, natural theology; with a low threshold for claiming scientific proof, Paley allowed his love for his Creator to color all that he could see. It's clear from his writings that Darwin was in love with the same natural world as Paley, but wanted to know more about how it worked (Darwin 1859, pp. 489–90). Nor was religious faith—whether wrapped up in the design argument or not—left behind in the Enlightenment. Numerous devout scientists practice their religions and pursue their science without conflict. Likewise, many religious leaders and organizations have formally declared their support for modern science, including evolution, and for the scientific study of evolution and Earth history (Matsumura 1995; K. B. Miller 2003; K. R. Miller 1999; Roughgarden 2006; Wright 2003). Earth science educators need to teach Earth science as science, to be clear about what science is and what it isn't, and to hope that this understanding grows into the consciousness of new generations of lawmakers, school board members, parents, and teachers.

About a year before my intelligent design discussion with a tow-truck driver in Seattle, I was being chauffeured in a VW dealer courtesy car back to my hotel in Mesa, Arizona. The affable driver informed me how wonderful it was that the Grand Canyon had been formed essentially overnight during Noah's Flood just a few thousand years ago. He reported that he had recently learned this lesson at an evening class held by his church. I wasn't about to contradict him: he was driving. But I was thankful that he had received his Flood geology lessons at his church, rather than in public school. Furthermore, I shared in his wonder at the glory of the canyon. Placing my hand on the Vishnu schist at the bottom of the Grand Canyon decades ago had been an awe-inspiring, religious experience for me—made all the more awesome by the insights gained by Lyell, Darwin, and the many scientists who followed, developing ever-sharper and older estimates of the true age and dynamic processes of the Earth.

REFERENCES

Allen, W. P., and J. Witt. 2004. *The Privileged Planet*. DVD. Illustra Media, La Habra, CA.

Brauer, M. J., B. Forrest, and S. G. Grey. 2005. Is It Science Yet? Intelligent Design Creationism and the Constitution. *Washington University Law Quarterly* 83(1): 1–149.

Bronowski, J. 1973. *The Ascent of Man*. Little, Brown, Boston, MA.

Brooke, J. H. 1991. *Science and Religion: Some Historical Perspectives*. Cambridge University Press, Cambridge, UK.

Browne, J. 1995. *Charles Darwin: Voyaging*. Princeton University Press, Princeton, NJ.

Buckland, W. 1836. *Geology and Mineralogy Considered with Reference to Natural Theology*, vol. 1. Bridgewater Treatises. William Pickering, London.

Center for the Renewal of Science and Culture. The Wedge Strategy. www.antievolution.org/features/wedge.html.

Darwin, C. 1836. *The Voyage of the Beagle*. Repr., 1972, Bantam Books, New York.

———. 1859. *On the Origin of Species*. Repr., 1967, Atheneum, New York.

Dean, C. 2007. Believing Scripture but Playing by Science's Rules. *New York Times*, February 12.

Discovery Institute. 2006. The "Wedge Document": So What? February 3. www.discovery.org/scripts/viewDB/filesDB-download.php?id=349.

Downey, R. 2006. Discovery's Creation. *Seattle Weekly*, February 1, pp. 19–26.

Durant, W., and A. Durant. 1965. *The Age of Voltaire*. Simon and Schuster, New York.

Feiss, G. 2007. Letter to NAGT Members Concerning May 2005 Issue of *JGE* Entitled "Who Believes What?" *NAGT Annual Report: 2004–2005*, March 18, pp. 5–7. www.nagt.org/files/nagt/organization/nagt-ar05.pdf.

Forrest, B., and P. R. Gross. 2004. *Creationism's Trojan Horse: The Wedge of Intelligent Design*. Oxford University Press, New York.

Gonzalez, G., and J. W. Richards. 2004. *The Privileged Planet: How Our Place in the Cosmos Is Designed for Discovery*. Regnery, Washington, DC.

Humes, E. 2007. *Monkey Girl: Evolution, Education, Religion, and the Battle for America's Soul*. Harper Collins, New York.

Killen, P. O. 2003. Religious Futures in the None Zone. Pp. 169–84 in P. O. Killen and M. Silk, eds., *Religion and Public Life in the Pacific Northwest: The None Zone*. Rowman and Littlefield, Walnut Creek, CA.

Kubrick, S., and A. C. Clarke. 1968. *2001: A Space Odyssey*. Repr., 2000, Turner Entertainment Company. www.kubrick2001.com.

Liberty University. 2007. Doctrinal Statement. March 18. www.liberty.edu/index.cfm?PID=6907.

Lyell, C. 1830. *Principles of Geology*, vol. 1. Repr., 1990, University of Chicago Press, Chicago, IL.

Matsumura, M. 1995. *Voices for Evolution*. National Center for Science Education, Berkeley, CA.

McLuhan, M. 1964. *Understanding Media: The Extensions of Man*. Repr., 2003, Gingko, Corte Madera, CA.

Miller, K. B., ed. 2003. *Perspectives on an Evolving Creation*. Eerdmans, Grand Rapids, MI.

Miller, K. R. 1999. *Finding Darwin's God: A Scientist's Search for Common Ground between God and Evolution*. Harper Collins, New York.

Miller, K., and B. Stein. 2008. *Expelled. No Intelligence Allowed*. Promise Media Corporation, Santa Fe, NM. www.expelledthemovie.com.

Nelson, P. A. 2002. Life in the Big Tent: Traditional Creationism and the Intelligent Design Community. *Christian Research Journal* 24(4). www.equip.org/free/DL303.htm.

Numbers, R. L. 2006. *The Creationists: From Scientific Creationism to Intelligent Design*, exp. ed. Harvard University Press, Cambridge, MA.

Paley, W. 1802. *Natural Theology; or, Evidences of the Existence and Attributes of the Deity Collected from the Appearances of Nature*, illus. ed. Repr., 2003, Kessinger, Whitefish, MT.

Pennock, R. T. 2000. *Tower of Babel: The Evidence against the New Creationism*. MIT Press, Cambridge, MA.

———, ed. 2001. *Intelligent Design Creationism and Its Critics: Philosophical, Theological, and Scientific Perspectives*. MIT Press, Cambridge, MA.

Ross, M. R. 2003. Intelligent Design and Young-Earth Creationism: Investigating Nested Hierarchies of Philosophy and Belief. Abstract, paper no. 244-4. Geological Society of America, annual meeting, Seattle, WA.

———. 2005. Who Believes What? Clearing Up the Confusion over Intelligent Design and Young-Earth Creationism. *Journal of Geoscience Education* 53(3): 319–23.

Ross, M., and P. Nelson. 2006. A Taxonomy of Teleology: Phillip Johnson, the Intelligent Design Community and Young-Earth Creationism. Pp. 261–75 in W. A. Dembski, ed., *Darwin's Nemesis: Phillip Johnson and the Intelligent Design Movement*. InterVarsity, Downer's Grove, IL.

Roughgarden, J. 2006. *Evolution and Christian Faith: Reflections of an Evolutionary Biologist*. Island, Washington, DC.

Scigliano, E. 2006. The Evolution of Bruce Chapman: How Seattle Became Home to Intelligent Design. *Seattle Metropolitan*, March, pp. 68–76.

Terry, M. 2004. One Nation, under the Designer. *Phi Delta Kappan* 86(4): 264–70. www.pdkintl.org/kappan/k_v86/k0412ter.htm.

———. 2005a. Art and Evolution. *The Science Teacher* 72(1): 22–25.

———. 2005b. Intelligent Design, or Not: Dr. Strangescience, or How I Learned to Stop Worrying and Love the Wedge. *New Horizons for Learning Online Journal* 11(3). www.newhorizons.org/trans/terry.htm.

———. 2005c. Putting Paley in His Place: History as an Ally in the Teaching of Evolution. *Geological Society of America 2005 Annual Meeting*, session 6. Abstract and recording. October 16. http://gsa.confex.com/gsa/2005AM/finalprogram/abstract_91525.htm.

———. 2005d. Tending the Tree of Life in the High School Garden. Pp. 78–82 in J. Cracraft and R. Bybee, eds., *Evolutionary Science and Society: Educating a New Generation*. BSCS/AIBS, Colorado Springs, CO.

———. 2007. What's Design Got to Do with It? *Independent School* 66(2): 40–47. www.nais.org/ismagazinearticlePrint.cfm?print=Y&ItemNumber=149392.

Terry, M., and S. Linneman. 2003. Watching the Wedge: How the Discovery Institute Seeks to Change the Teaching of Science. *Washington State Science Teachers' Journal* 43:12–15.

Twain, M. 1889. *A Connecticut Yankee in King Arthur's Court*. Repr., 1960, Washington Square, New York.

Voltaire. 1756. On the Lisbon Disaster, or an Examination of the Axiom "All Is Well." Repr., 1965, pp. 721–22 in W. Durant and A. Durant, *The Age of Voltaire*. Simon and Schuster, New York.

———. 1759. *Candide, or Optimism*. Repr., 1981, Penguin Books, New York.

White, T. H. 1954. *The Bestiary: A Book of Beasts, Being a Translation from a Latin Bestiary of the Twelfth Century*. Repr., 1960, G. P. Putnam's Sons, New York.

Wright, R. T. 2003. *Biology through the Eyes of Faith*. Harper Collins, New York.

SIX · It's Not about the Evidence

The Role of Metaphysics in the Debate

CHARLES E. MITCHELL

INTRODUCTION

Scientific and religious accounts of human origins are founded on very different philosophical approaches to knowledge. In particular, the apparent conflicts between science and religion with respect to human evolution owe their origin to the different roles that metaphysics plays in the two approaches to knowledge and to understanding what it means to be human. At their most divergent, these perspectives place humans in a very different place within their explanatory framework and define what it means to be human in very different ways. For many evolutionists, humans are merely one of a vast number of well-adapted but fundamentally natural outcomes produced by an unguided process (natural selection), whereas, for many religious people, humans are the central reason for being of a divinely created and actively sustained, goal-directed universe. These differences in basic metaphysical propositions inevitably color the work that the two groups do. The tension between these different viewpoints underlies the continuing reticence of many people to accept the scientific account of human origins.

In this essay, I offer my perspective as an Earth scientist in general and a paleontologist in particular. First, I summarize some features of the intelligent design (ID) argument and its connection to previous views about the relationship between Christian theology and Earth history. I aim to examine the question of why this discussion persists. Given the deep understanding that scientists have gained about the nature of our origins, why is it difficult for so many people to accept the fact of evolution? The basic ideas pushed by ID advocates have been around since before Darwin's time. They have gotten no better; they continue to be soundly rejected by scientific study, and yet they linger. This persistent resistance relates to differences in the way we approach knowledge and the types of conclusions to which those differences lead. Such philosophical differences are difficult to resolve because they hinge on presuppositions that are indispensable to the scientific and religious approaches to knowledge.[1] Thus, we must accept these philosophical differences and consider how to deal with them. Finally, I will comment on the problems that arise when we fail to appreciate the incommensurability of creationist and evolutionist approaches, explain why we must contemplate this risk, and suggest a path that leads to greater understanding.

INTELLIGENT DESIGN, HUMANS, AND PURPOSE

The intelligent design movement is the current incarnation of the creationist response to scientific accounts of human evolution. In his book *Darwin's Black Box*, ID proponent Michael Behe (1996) argues that evolution appears to make sense until we look at the underlying mechanisms at a molecular level. Behe (who is a molecular biologist) contends that when we unpack the elegant machinery that is living organisms, we find that they're ferociously complicated things, with innumerable and essential molecular machines inside them that are "irreducibly complex." What he means by irreducible complexity is that if any single component of the mechanism is omitted, it will not function; the machine cannot be reduced to separate parts that can be added one by one to make the functioning mechanism. And if the machines can't be broken down to their component parts, they can not have evolved because evolution is a piece-wise, gradual process in which parts are added bit by bit. Take a household mousetrap, for instance. Behe points out that if any one of the parts is missing, the trap does not function. True enough. Our understanding of the evolutionary process, however, indicates that this is not how adaptation comes about. Darwin (1859) confronted this very same argument from his contemporary St. George Mivart and effectively answered it in the chapter "Difficulties of the Theory," which he added to revisions of *On the Origin of Species*. Substantive critiques of Behe's arguments are many (see, for instance, Kitcher 2001; Brauer and Brumbaugh 2001).

Behe is unconvinced by Darwin's suggestions that naturalistic evolution can construct new and essential functions by reshaping features originally acquired for other functions or by enhancing an originally minor advantage of a new feature until it becomes essential. Behe claims not that these mechanisms are impossible but simply that they are highly unlikely. William Dembski's (1998) application of information theory to challenge the plausibility of adaptive evolution follows a similar trajectory, albeit in different terms. The ID proponents find these notions unconvincing not because the data are poor or because the theory is inadequate, but because as ID proponents they have a prior commitment to the existence of purpose in the natural world.[2] If we begin the study of evolution with a conviction that it is leading to a particular outcome (humans, say), we will naturally find explanations that omit purpose to be unsatisfying—indeed, they may appear to be a direct challenge to our faith. The philosopher of religion Alvin Plantinga arrives at precisely this conclusion: "According to Scripture, God has often treated what he has made in a way different from the way in which he ordinarily treats it; there is therefore no initial edge to the idea that he would be more likely to have created life in all its variety in the broadly deistic way.[3] In fact it looks to me as if there is an initial probability on the other side; it is a bit more probable, before we look at the scientific evidence, that the Lord created life and some of its forms—in particular human life—specially" (Plantinga 1991, reprinted in Pennock 2001, p. 130). Plantinga goes on to conclude that, given what he knows from faith and his assessment of the scientific evidence, the claim that humans evolved naturally is doubtful and that a new approach to understanding human origins is needed: "'Unnatural Science,' 'Creation Science,' 'Theistic Science,'—call it what you will: what we need when we want to know how to think about the origin and development of contemporary life is what is most plausible from a Christian point of view. What we need is a scientific account of life that isn't restricted by that methodological naturalism" (Plantinga 1991, reprinted in Pennock 2001, p. 139).

To the untrained eye the claims of ID proponents such as Behe and Dembski appear to be scientific. They employ the language of science to talk about the difficulty of devising intermediary steps by which to evolve the critical functions of molecular structures (Behe 1996) or the mathematical improbability of evolving complex features by chance (Dembski 1998). If the arguments stuck purely to such issues, the debate would be both scientific and over, since these few scientific claims are quite weak. However, ID proponents cannot stick to purely scientific issues because their purpose is religious, not scientific. Their goal is to promote specific Christian beliefs—ideas that cannot be questioned.[4] Intelligent design arguments inevitably take the form of religious apologetics, with their goal of finding in the natural world signs of the existence of what they know a priori to be true. Their arguments stem from philosophical principles and articles of faith.

CAUSE AND MEANING

The basic philosophical contrast between science and ID apologetics that I wish to emphasize here has to do with attitudes about causation; in particular, how each approach regards the role of meaning and purpose as an explanatory agency in the natural world. By *meaning* I refer here to the value that we attach to objects, experiences, actions, and ideas. Meaning in this sense is intimately connected to notions of purpose, and by purpose I have in mind the usual sense of the intended value of a thing—not only that an object or notion has some meaning but also that this value was what it was intended to have. The purpose of a hammer is to pound nails. This is what it was designed to do. Design, then, is the process of conceiving a purpose and bringing it into existence. We use these words in a similar way when we say that we seek to add meaning and purpose to our life. We seek to discover some pursuit to which we attach a positive value and this provides us with purpose: the motivation to pursue our chosen course with the intention to achieve the associated value.

When we think about objects in the world, even a fairly ordinary object, a chair for instance, we can understand that chair in different ways and from varying perspectives. We might ask, what is the cause of a chair? Where do chairs come from and why do they possess the properties that they do? One useful way to organize our thinking about the nature of causes—the how and why of chairs, for instance—originated with Aristotle, who described cause in terms of four components: *formal cause*, *material cause*, *efficient cause*, and *final cause*. With regard to a chair, its *formal cause* is the laws of physics that govern all matter. The *material cause* is the matter out of which the chair is made, whether springs and metal, or wood. The *efficient cause* of the chair is the building process itself—the actions of those who built the chair. Aristotle's *final cause* concerns meaning and purpose. This category of causality is the crux of the debate between creationists and evolutionary scientists.

Some aspects of a chair we can know fairly well. We know that a particular chair is manufactured of specific materials; we can ascertain these empirical properties with confidence. It's not as if there aren't any questions about how we understand these properties or how well we might measure them, but we generally can make a pretty precise assessment of the material properties of the chair. What about the notion of this object *as a chair?* What *is* a chair exactly? This question strikes us as strange simply because the answer is so obvious. A chair is something people sit on. It's not just an object made from metal and springs, for example. It has a purpose, and so the meaning of a chair is defined by its purpose. The problem with this answer is not that it is an arcane notion, but that it depends on context. The meaning of a chair to a honeybee is quite different than it is to a person since bees don't sit on chairs. To a honeybee, a chair is just an obstruction to its flight through a room. With respect to a neutrino, a chair doesn't exist at all. What is to us a solid object is mostly empty space when experienced at the atomic scale by a massless, chargeless, subatomic particle moving at the speed of light. So meaning and even, to some degree, physical properties like "there-ness" are a function of context. *The chair is not either solid or not-solid; it is both solid and not-solid simultaneously, and whether it appears to us as one or the other is a product of our perspective.* The unitary reality of the chair, as we perceive it, is an illusion—a product of our capacities as observers, not a product of the thing itself. The meaning of the chair to us as observers is a function of our metaphysics—of the set of beliefs that we bring as we examine the chair. Its meaning is a function of our sense of what's important in the world, what the purposes of things are, what the ultimate nature of existence is. Meaning is not inherent in an object.

Our ideas of chairs, of mountains, of the Earth, indeed all our knowledge is in part a function of us. It's as if we are trying to look out at the world through a window, but everything we wish to see is colored by the reflection of ourselves on the glass. John Steinbeck examined this issue in his novel *Cannery Row:*

> On the black earth on which the ice plants bloomed, hundreds of black stinkbugs crawled. And many of them stuck their tails up in the air. "Look at all them stink bugs," Hazel remarked, grateful to the bugs for being there.
>
> "They are interesting," said Doc.
>
> "Well, what they got their asses up in the air for?"
>
> Doc rolled up his wool socks and put them in the rubber boots and from his pocket he brought out dry socks and a pair of thin moccasins. "I don't know why," he said. "I looked them up recently—they're very common animals and one of the commonest things they do is put their tails up in the air. And in all the books there isn't one mention of the fact that they put their tails up in the air or why."
>
> Hazel turned one of the stinkbugs over with the toe of his wet tennis shoe and the shining black beetle strove madly with floundering legs to get upright again. "Well, why do *you* think they do it?"

"I think they are praying," said Doc.

"What!" Hazel was shocked.

"The remarkable thing," said Doc, "isn't that they put their tails up in the air—the really incredibly remarkable thing is that we find it remarkable. We can only use ourselves as yardsticks. If we did something as inexplicable and strange we'd probably be praying—so maybe they are praying."

"Let's get the hell out of here," said Hazel. (1945, pp. 37–38)

Science is a particularly effective tool for disentangling what we bring to our understanding of the world from what exists independent of human perception. Science gains knowledge of the world by recognizing the problem posed by our limited perspective and by seeking to reduce or control the effects of this projected self. To do this we rely entirely on natural causes that can be studied empirically and we apply a set of procedures that limit the opportunity for our hopes and fears to intrude on the analysis. This approach to understanding, which focuses on the first three of Aristotle's four causes, is commonly referred to as *methodological naturalism* (see, for example, chapter 7 for a more complete discussion of methodological naturalism).

For many questions having to do with natural events, particularly those involving inanimate phenomena, it is possible and entirely noncontroversial to analyze their causes without including final cause. When we consider the history of the Earth, we might talk about why there are mountains in an efficient cause sense of plate tectonics and subduction and things of this sort, or we might talk about mountains in the materials property sense of how strong continents are and how tall mountains can be in consequence. But we typically don't talk about the *purpose* of mountains; effects perhaps, but not purpose.

Even with familiar objects that have a purpose obvious to us, we may usefully take this tack. We can accept that there may or may not be a final cause for chairs, set that question aside, and analyze from a purely naturalistic perspective what they are made of, how stable they are, how well they move around the floor. The designer's intentions are not relevant to these questions. But when we want to ask questions about what value the object has as a chair, we must take account of meaning and purpose. In the case of chairs, their purpose dictates the other choices about material and process.

Science strongly prefers to focus on the first three of the Aristotelian causes—law, material, and process—because these aspects of cause are relatively objective.

They're not entirely objective, but they're relatively objective compared to questions of purpose, such as why stinkbugs might put their tails in the air. This question, as Steinbeck points out in the discussion between Doc and Hazel, is highly subjective. We can easily devise an instrument to measure the weight and strength of a chair, but it is quite difficult to assess its purpose, except, as Steinbeck says, by gauging it relative to what we might intend its purpose to be. Final cause just isn't accessible to the same degree that these first three causes are.

In the case of living organisms, however, purpose moves front and center in our concerns. The features of organisms appear to us to have purpose: wings, teeth, hooves, all are clearly well suited to serve a specific function in a specific circumstance. How did these features come about and does their cause involve something we can reasonably call purpose? Christian theologians (e.g., Paley 1809) have argued that the evident good design of organisms was clearly a product of God's will and, furthermore, the evident design was itself evidence of the wisdom and benevolence of God. ID takes up this argument in modern terms and claims that the high degree of complexity of the functional features that organisms possess is such that they can be explained only by some form of purposeful process—by the actions of a creator. In contrast, scientists, relying on natural causes and seeking to understand the phenomena of adaptation on their own terms rather than in the light of their significance for our search for meaning in the universe, have come to the conclusion that the fit of organism to environment is an unguided outcome of the struggle for existence, that is, a product of natural selection.

In *On the Origin of Species* (1859), Darwin dispensed with intention and *inverted the relationship between cause and effect in design.* Natural selection does not result in purposeful, forward-thinking design. Instead, it is an incidental effect of genetic differences among the members of a local population, a population of rabbits, say, that leads to the differential survival or reproduction of some of these individual rabbits relative to others. Increased speed or camouflage is a fortunate outcome of the mutation in a particular context, not its purpose. The mutation is retained and spread through a population because it has this beneficial effect—because it adds constructively to the other beneficial changes in the constitution of rabbits that have preceded this mutation. The net result is a successfully functioning rabbit well suited to its current circumstances. Natural selection provides a natural explanation for the apparent design—an explanation in which purpose is not necessary. Darwin's evolutionary mechanism, with its emphasis on natural causes, therefore, substitutes directly for the hand of God. As I will discuss later, the relationship between God and natural selection may be more complex than this, but the situation is commonly understood in this either-or form.

PROGRESS AND DIRECTIONALITY

For several days, a young man is dragged by his parents from one Civil War battlefield site to another. Finally, unable to contain his curiosity any longer, he asks, "Why were all the Civil War battles fought in national parks?" The notion that the present state is something that historical change was seeking in a purposeful way leads us to invert the connection between cause and effect and, in this joke, to reverse the historical priority of Civil War battles and national parks. Creationists make precisely the same mistake. The assumption of the centrality of humans in the natural order of the universe is intrinsic to many religious beliefs and is a powerful notion that is inherently contrary to the essential skeptical character of scientific epistemology. It leads us to expect change to lead from some earlier, inferior state to a superior, later state, namely, us. In a recent cartoon, the political satirist Tom Toles parodies this mistake: a teacher standing in front of a blackboard with the words "Biology—Today's Debate" asks his students, "If humans evolved from monkeys, why are there still monkeys?" while in an adjacent room a history teacher asks her class, "If Americans came from English Pilgrims, why are there still Englishmen?" (*Washington Post* 2005).

I claim not that Earth history and human evolution are not progressive but only that we cannot accurately study these histories by starting with that assumption. Holding this belief is precisely the error that underlies Behe's misunderstanding of evolution: he notes that the current function of some feature (a flagellum, for instance) cannot be constructed piece-wise because it will not function without all of its pieces, as if this current function had been what evolution was aiming at. Theistic science is junk science. I understand Behe's desire to see the spiritual truth of the Bible taken seriously, but what he advocates is not a productive way to go about it.

Notions of progress, like purpose, are inherently subjective. Just as is the case with the other topics examined so far, there has been a long history of discussion about how to define progress and whether or not the history of life or the fossil record exhibits evidence of progress (e.g., Nitecki 1988; Ruse 1996; Rosslenbroich 2006). The evolutionist and vertebrate paleontologist George Gaylord Simpson (1952) discussed concepts of progress in his book *The Meaning of Evolution*. The book provides a clear and general treatment of the difficulties of defining progress and a thorough look at the case for whether or not the fossil record supports a belief that progress is a general feature of the natural world or, more locally, of human evolution. His answer was no, progress is not a general or inherent tendency of evolution, but yes, human evolution does exhibit a form of progress. Before I explain why I think this is a reasonable but not very helpful answer, let's update the argument just a bit.

In their early incarnations, progressivist views held that organic existence is organized along an ordered sequence from the lowest, least sophisticated organisms (nearly formless, primordial slime, in essence) to the most advanced, highest organisms, which are humans. Such notions of general progress—of an inherent striving toward perfection—are intimately connected to the origin of evolutionary theories in the work of Lamarck, Chambers, and Erasmus Darwin (Bowler 1984; Ruse 1996). Although notions of general progress as a universal perfecting force, including notions of human preeminence or inevitability, remain prominent in the writing of many leading evolutionists during the early half of the nineteenth century (e.g., Fisher 1930, Dobzhanski 1962, Huxley 1942), writers in the latter half, such as Simpson (1952), Williams (1966), Dawkins (1986), Gould (1988), and Provine (1988), reject this metaphysical inclination, which has since largely (but not entirely) disappeared from current evolutionary thought (see discussions in Ruse 1988, 1996; McShea 1994; Rosslenbroich 2006).

A more limited type of general progress does persist in the form of a range of empirical claims about evolutionary trends, however. In terms of claims based on the fossil record, two are particularly noteworthy. Raup (1988) pointed out that in contrast to notions of "improvement" or "increased complexity" that typically underlie claims for progress exhibited by the fossil record, rates of extinction within lineages can be measured objectively, and uncertainty in the measure can be quantified as well. Based on data for genera and families of marine shallow-water invertebrates, Raup suggested that there has been a general decline in extinction rates within classes over the past 500 million years. This trend may indicate that there has been a general selection against lineages that are prone to go extinct and that the result is the accumulation of more extinction-resistant, slower-evolving lineages (the data exhibit a matched decline in the rate of appearance of new genera and families as well). Now, this outcome is hardly the sort of march up the ladder of betterment toward humans that previous notions of progress entailed, but it might be a form of progress nonetheless. What is less clear is whether the pattern is real or whether it is an artifact of how we paleontologists treat our data—of how we form the groups we call genera and families given the bits and pieces that remain in the fossil record. The ancient record is much less complete than the more recent record, which is itself less well known than is the living fauna. As Raup (1988) noted in this and several other contexts, comparisons across this range of ages and inherent biases may yield trends in extinction and origination rates that are largely a product of the biases themselves rather than the original patterns of evolution. More recent studies of the diversity and evolutionary volatility of lineages tend to confirm the basic pattern that Raup described (e.g., Bambach, Knoll, and Wang 2004), but the jury is still out on the degree to which the pattern is real or artificial (compare, for instance, the conclusions of Crampton et al. 2006 and Kowalewski et al. 2006).

Recent work by Wagner, Kosnik, and Lidgard (2006) adds a different twist to this argument from fossil diversity data. Their analyses suggest that the ecological structure of communities has become strikingly more complex following the late Permian mass extinction. The Mesozoic and Cenozoic communities also experienced a major evolutionary expansion in the diversity of the invertebrate groups that today dominate marine ecosystems. This coincidence suggests that these "modern" invertebrate groups are in some way more interdependent or more specialized, or both, than their more primitive, Paleozoic counterparts.

Gould (1988, 1996) presented a quite different argument: that the whole notion of progress was so freighted with our hopes and fears as to be entirely inaccessible as an empirical phenomenon or, worse still, that it has inevitably become so contaminated by cultural bias, racism, and hatred that the notion of progress is best avoided entirely. Instead, he argued, we might ask fruitfully whether evolutionary change exhibits directionality. Does it generally seem headed in some particular direction or do individual lineages exhibit patterns of sustained transformation through time? This is an empirical question that perhaps can be tackled objectively, and many authors have picked up this challenge.

In *Full House* (1996), Gould expanded this theme by arguing that the appearance of progress within the fossil record arises largely from the directionality of transformative processes themselves: that they must begin at the beginning and can only move on from this point of origin. Life, Gould pointed out, began with simple, small organisms, and from there has expanded into a broad array of organisms with a wider range of sizes, forms, and ways of life. Some of these newer animals and plants are larger or more complex, or more specialized, and so on, than their ancestors, and some groups have expanded the range of occupied habitats (moving onto land and into the air, for instance), but all of this, he argued, was simply an inevitable consequence of the minimalist state in which life began.

Given that life began with a small organism with a narrow range of capacities (and any individual organism must be limited in this way), evolution simply expands outward toward the unoccupied range of possibilities that surround the starting point. No other direction of change is possible. But once fully complex (or larger, or more specialized, or whatever) organisms evolve, then directionality as a general feature disappears and evolution follows whatever path (to more or to less complexity, say) is advantageous at that moment, given the particular local conditions that each species experiences. Some of these unique paths may exhibit directionality, but this is a purely local directionality with no general tendency toward complexity, or specialization, or even diversification.

My own work on the evolution of graptolites provides an example of local directional evolution. Graptolites are a well-studied group of extinct animals. They were members of the suite of planktic organisms that inhabited the near-surface waters of the open ocean from about 490 to 410 million years ago (from the beginning of the Early Ordovician into the Early Devonian epoch, in the formal terminology of the geological time scale). Graptolites constructed skeletons containing a series of tubes, each of which housed an individual animal. These animals were each formed by asexual budding from the founder individual, which was itself probably the product of a fertilized egg—hence they were colonial animals like corals and bryozoans. Because they became extinct in the Devonian and because very little of their internal anatomy is preserved in any of the available fossils, there is a great deal about their basic biology that is entirely mysterious. But like all good mysteries we have many tantalizing clues to puzzle over.

Among the things we can determine from fossilized graptolites is that the shape of their colonial skeleton influenced the ways in which graptolite animals fed and how the colonies moved through the water to find food and mates. Graptolite colonies had very regular forms that were precisely repeated in each member of their populations. This regularity is comparable to the amount of variability we find in the functionally important features of living organisms—the teeth of beavers and the claws of cats, for instance. Studies of the way that fluids flow around models of graptolite colonies support this general sense that the colony shapes had functional importance (e.g., Rigby and Rickards 1989; Rigby 2003). Accordingly, it is reasonable to think that evolutionary changes in colony form that are revealed by the fossil record of graptolites correspond to changes in the way that the colonies functioned.

In nearly all graptolites the dwelling tubes had thin, flexible walls composed of organic material, probably collagen. The tubes grew by the addition of new wall increments to the open end of the tubes—a bit like constructing a tower from bricks, working upward from its foundation. The growth increments of the tube walls are visible on the fossils of the colonies as faint growth lines that cross the tubes. In many graptolites the dwelling tubes are shaped like simple cones or pipes, but in others the tubes are curved or have other more complex shapes, and some graptolites exhibit spines that projected out from the tube wall on one or more of the tubes.

In one particular group of graptolites, the dwelling tubes all possess a forked set of spines that formed near the open end of the tube. These forked spines became linked together by vertical, threadlike strands of skeletal material in a few of these species. From one of these somewhat unusual graptolites, an elaborate sequence of new species evolved during the Late Ordovician. Over a period of about 5 million years the colony form constructed by graptolites in this lineage became drastically altered from the ancestral condition with complete tube walls to a form in which the tubes developed interconnecting thickenings along their outlines, and the material in the wall became thinner and thinner until it was entirely lost. The tubes became scaffoldlike, mere outlines of colonies. The first pair of tubes (one on each side) and the central cone (which housed the founding individual) still possessed dense walls, but the later tubes are represented only by a series of interconnected rods of skeletal material. Forked spines linked the scaffolding (which outlines the dwelling tubes) to the enclosing mesh of threads. In many species in this lineage, the outer set of spines and linked threads became more numerous and more complex until they completely surrounded the colony in a basketwork mesh. At the same time even the earliest tubes became scaffoldlike. The fossils of these graptolite colonies resemble small fuzz balls: they look, for all the world, like fossil dust bunnies!

The evolutionary history of these species reveals that the scaffolds and fuzz balls are in some sense more advanced graptolites that evolved from the more ordinary forms with complete tube walls. This development is suggested in figure 6.1 by the set of brackets that link closely related species into a set of nested groups, thereby forming a sort of evolutionary tree. In the technical language of my field this "tree" is a cladogram and it depicts the sister-group relations of species based on a computer-assisted analysis: a data set that describes their colony form and

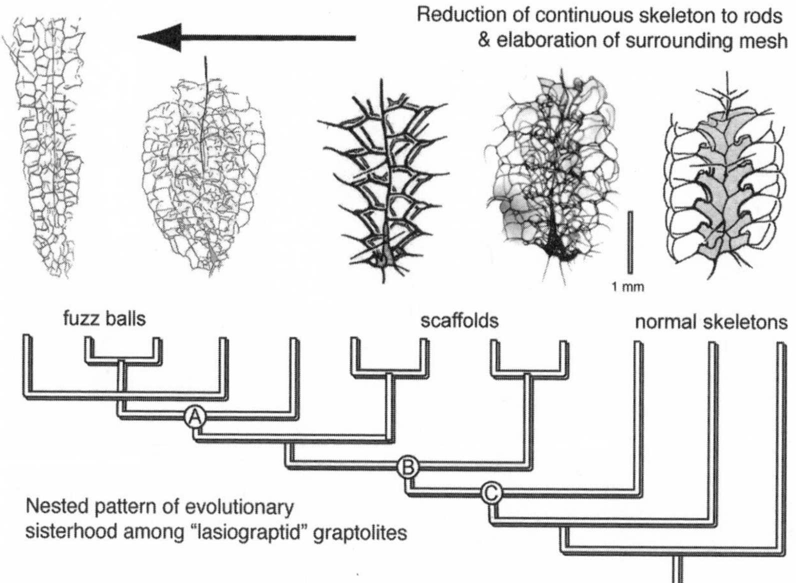

FIGURE 6.1
Cladogram that shows the sequence of branching events that occurred during the evolution of this group of retiolite-type (lasiograptid) graptolites. Branching order moves upward through the tree. Lower branching points (such as point C) correspond to more ancient speciation events (divisions of once-continuous populations into two separate species), and higher branches (as at point A) to later events. During the course of the evolution of this lineage, the graptolite colonial form changed from that shown on the right (normal skeletons) to highly modified forms (fuzz balls) (illustration by author).

patterns of growth. It does not directly depict the relative age of the several species or ancestor-descendant relations (who gave rise to whom). The more evolutionarily "advanced" fuzz ball species are those on the left and they share a single common ancestor, which we can think of as represented by the place on the tree where the five fuzz ball branches link together (see point A in figure 6.1). Working our way down the tree, we can tell they share an earlier common ancestor at point B with the several closely related scaffolds, and this set is together an evolutionary descendant of a species at point C that is the common ancestor of all these odd

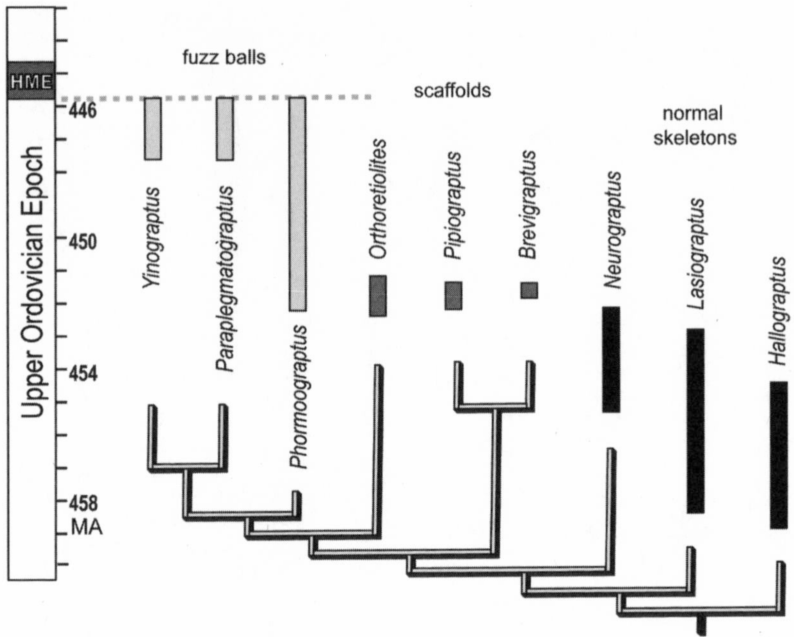

FIGURE 6.2
Diagram showing the correspondence between the sequence of evolutionary origin of the retiolite-type genera and their known time of occurrence in the fossil record. More archaic forms, judged on the basis of their morphology, occur in older rocks more often than the forms that appear to be increasingly modified. These archaic forms were evidently successively replaced by more advanced types, displaying the sort of directional change that matches our expectations for a progressive evolutionary trend. The entire group, however, went extinct during the Hirnantian Mass Extinction (HME), near the end of the Ordovician Period (illustration by author).

graptolites and some of their more normal cousins. Thus, this set of graptolites exhibits a strongly directional change in colony form.

Not only is the history of evolution in this group of graptolites strongly directional in terms of the change in overall colony design; it is also directional in time (see figure 6.2). The scaffolds appear in the fossil record after the species with normal skeletons and forked spines that are their likely ancestors, and the scaffold-like species subsequently diversified as the older forms went extinct. The fuzz balls are younger still and, like the scaffolds before them, they replaced their predecessors. As new forms evolved, the earlier ones disappeared and the range of colony forms present in the lineage shifted continuously with time, away from the standard graptolite form to this odd new type—the retiolite-type colony. This outcome is an evolutionary trend. In particular, it is what has been called an active or driven trend (McShea 1994) and it is evidence of response to a steady evolutionary pressure to evolve along a particular adaptive path.

Interestingly, this is not the only time that retiolite-type colonies evolved among my favorite animals. It also happened at least three other times during the Ordovician Period. Each time the trend started from a different lineage and resulted in distinctly differently constructed retiolites. The most successful of these retiolite-type graptolite lineages evolved during the Silurian Period (the next interval of Earth history after the Ordovician Period). This lineage produced a larger number of species and lasted a great deal longer than any of the four times other graptolites followed this path during the Ordovician (including the lineage discussed earlier). The situation is analogous to the evolution of flying animals. Insects, pterosaurs, bats, and birds have each given it a try; each began from a substantially different starting point and each solved the problems inherent in flight in different but similar ways. The evolutionary convergence in these forms is a testament to the singular effects of history and the unifying effects of the physics of flight. So, too, the development of retiolite-type graptolites: they evolved in a progressive way, each time converging on a common solution to the problem of being a graptolite with a reduced skeleton.

In each case, however, the retiolite lineages were relatively limited in the number of species that evolved and each trend terminated in total extinction. The most diverse and long lasting of the Ordovician retiolite-type lineages, the one I described earlier, was driven to extinction during the great mass extinction that took place at the end of the Ordovician (Chen et al. 2005). This was one of the largest mass extinction episodes known—larger than the one at the end of the Cretaceous, when dinosaurs, ammonites, and a host of other species went extinct, and second only to the great Permian mass extinction (Bambach, Knoll, and Wang 2004). The limited success of retiolite-type graptolite lineages suggests that the particular mode of life that their peculiar morphology was built to foster was a limited, specialized sort of opportunity, and was one that was subject to crashes during bad economic times, so to speak. Thus, although we can reasonably regard the evolution of retiolite-type graptolites as progressive, it's a very limited, local sort of progress. Many other graptolite lineages exhibit other limited directional trends and frequently show convergent evolution, in which previous adaptive solutions are reinvented by new lineages.

Graptolites are not unusual in the frequency or character of this limited progressive evolution. In fact, this example is quite typical. We can easily find cases of short-term, limited runs of progressive change, but they hardly ever add up to long-term change or lasting success. They all, like the various graptolite lineages, exhibit their own brand of success, with unique features that reflect their own particular histories and the special opportunities offered by the moment in time when they arrive at these junctures.

Earlier, I stated that Simpson (1952) concluded that there was no general or inherent progressive tendency within evolution (that is, no general path or type of change that can be found in all or even most lineages), but that indeed human evolution does exhibit a form of progress. We can now consider why I think this is a reasonable, but not very helpful, answer. First, from a purely empirical perspective, we must admit that the directional changes in human intelligence and social capacity (the features most commonly identified as the signatures of our progress) are at best local directional changes, just like the graptolite-driven trend I described earlier, and are not global features of evolution. Therefore, our particular triumph cannot be seen as the purpose toward which evolution in general has all along been working.

Secondly, our judgment that even these limited directional changes represent progress is deeply subjective and metaphysically bounded:

> The criterion natural to human nature is to identify progress as increasing approximation to man and to what man holds good. The criterion is valid and necessary as regards human history, although it carries the still larger obligation of making a defensible and responsible choice among the many and often conflicting things that men have held to be good. The criterion is also perfectly valid in application to evolution in general, provided we know what we are doing. Approximation to human status is a reasonable *human* criterion of progress, just as approximation to avian status would be a valid avian criterion or to protozoan status a valid protozoan criterion. It is merely stupid for a man to apologize for being a man or to feel, as with a sense of original sin, that an anthropocentric viewpoint in science or in other fields of thought is automatically wrong. It is, however, even more stupid, and even more common among mankind, to assume that this is the *only* criterion of progress and that it has a *general* validity in evolution and not merely a validity relative

to one only among a multitude of possible points of reference. (Simpson 1952, pp. 241–42; emphasis as in the original)

This is not to say that our progress is not real in some sense, but that, as Simpson notes, it's progress only from *our* perspective. Had they the capacity to express it, the drowning polar bears and thirsty giraffes that also inhabit this anthropogenically altered world might view our evolutionary history differently. And even if they were inclined to be generous, what we have attained would be valuable only in our context, not in theirs. Whatever tendency there is for evolution to pursue progressive paths, it does not lead in some general, universal direction, much less to us in particular.

This is what a consideration of the empirical record tells us. It is not an especially cheery set of insights in the sense that it does not encourage our inclination to search for benign purpose or meaning in the world. What, then, are we to conclude from this insight about the existence of meaning in nature?

CONFLICT BETWEEN SCIENCE AND RELIGION

Current controversy over the teaching of evolution and intelligent design is grounded in an inappropriate application of insights from science and religion to areas where they do not apply. Some Darwinians, for instance (e.g., Provine 1988; Dennett 1995; Dawkins 2006), have claimed that because it is possible to explain the effective design of living organisms by purely natural causes, no further explanation for purpose in the universe is needed. From this conclusion they infer that no other purpose or design exists (some equivocate here and say "most likely none exists"), and that therefore God does not (or probably does not) exist. Even if our theories are true, our scientific knowledge cannot exclude what it is not constructed to encompass, and scientific knowledge is constructed entirely within the domain of natural causes. People who believe that the world contains no ultimate purpose adopt their atheism because of some prior commitment and when they argue as I have sketched here they are simply using empirical knowledge as part of their metaphysical naturalist apologetics. It is possible that there is no ultimate purpose and that God does not exist, but this is not proved (or shown to be likely) by science. This claim may follow from a sound logical argument, but it is not science and has no business being touted as such.

The metaphysical naturalist position forces us to choose between abstract notions of Earth history and natural selection on the one hand, and our visceral and emotional experiences of spirituality on the other.[5] The scientific approach to knowledge draws on third-person narratives, whereas spiritual investigation is fundamentally a first-person experience. One of the most dire failings of the claim for a scientific justification for metaphysical naturalism is that it denies the validity of people's actual experiences. It is no wonder that so few people accept Darwinism in the form promoted by Dawkins (2006). While listening to the radio during the hubbub of the 2005 *Kitzmiller v. Dover Area School District* court case, in which this conflict most recently came to a head, I was struck by an interview with one of the students from the Dover high school. She supported teaching ID and, in response to the interviewer's question of why this was so, she said essentially this: that the notion of evolution was to her quite remote, whereas she daily felt God in her life. She knew that God was real and could not imagine rejecting this knowledge in favor of what she has been told evolution required of her. Whether her experience primarily reflects an objective external reality, a social convention, or an internal state of mind is an important issue about which we could have many interesting conversations, but in no case is the experience unreal. We need to find a way to accept such experiences as carrying an important truth of their own and yet also harmonize them with the valid knowledge that we gain from scientific study.

As I have already pointed out, some religious people, on the other hand, find the scientific account of human origins to be incredible. They reject this view not because the scientific evidence is poor or its theories incomplete (despite what they claim as the basis of their objections), but because they cannot understand, from the perspective of their strongly goal-directed theology, how apparently unguided processes could produce transcendent humans. From their prior commitment to a theistic universe they feel compelled to attack scientific knowledge as false or falsely presented, and so present a front of hostility to scientific understanding. This reaction also is inappropriate. The problem lies not with the scientific knowledge, but with a theology that demands empirical conformity with its spiritual insights. No amount of arguing over data or definitions of science will resolve this misunderstanding because its origin lies in the metaphysics, not the fossils or the evolutionary theory.

For me the strength of the spiritual perspective is that it seeks understanding not by trying, as science does, to minimize the coloring effect of our personal perspective, but by looking at the self directly to understand what it means to be human. It seeks to put the self in proper perspective relative to purpose and meaning and, for many, relative to God or some notion of the divine. Thus, science and spirituality serve very different purposes and hinge on very different underlying metaphysical propositions. They are not contradictory, but like the chair that is at once both solid and not-solid, science and spirituality have interconnected, albeit distinct, and simultaneous existences. Each has contexts in which it provides essential knowledge and contexts that lie outside its individual sphere of insight (Gould 1997, 1999; Clouser 2001; Ruse 2001, 2002).

How, then, might we proceed in an effort to gain all the benefits of our scientific and spiritual insights and to avoid the losses that conflict inflicts upon us? Like much else that I have discussed in this essay, this is a deep issue and one on which a great deal has been written. Nevertheless, I do think it worthwhile to point out a couple of possible alternative approaches that seem to be logically consistent and intellectually defensible.

One may incorporate natural selection and human evolution into one's religious understanding as the method by which the creator creates (see chapter 9). Many theologians and philosophers have taken this course (e.g., Pope John Paul II 1996; Peacocke 2001; Van Till 2001). Charles Darwin described natural selection as an ongoing, continuously engaged process: "It may metaphorically be said that natural selection is daily and hourly scrutinizing, throughout the world, the slightest variations; rejecting those that are bad, preserving and adding up all that are good; silently and insensibly working, *whenever and wherever opportunity offers*, at the improvement of each organic being in relation to its organic and inorganic conditions of life" (1859, p. 97, emphasis as in the original). This description comports well with some Christian theists' notions of a God who sustains and conserves the universe by direct engagement of his attention and who acts continuously in the world during the normal course of events, through natural processes:

> Science had pushed . . . God farther and farther away, and at the moment when it seemed as if He would be thrust out altogether, Darwinism appeared, and, under the guise of a foe, did the work of a friend. It has conferred upon philosophy and religion an inestimable benefit, by showing us that we must choose between two alternatives. Either God is everywhere present in nature, or He is nowhere. He cannot be here, and not there. He cannot delegate His power to demigods called "secondary causes." In Nature everything must be His work or nothing. We must frankly return to the Christian view of direct Divine agency, the immanence of Divine power in

nature from end to end, the belief in a God in Whom not only we, but all things have their being, or we must banish Him altogether. (Moore 1890, pp. 73–74)

According to the Bible, it is God, after all, who created the universe and, with it, the natural processes that govern its daily operation. Nothing about the process of natural selection itself requires a religious person to reject it as the hand of God any more than we are compelled by intellect to reject his role in the production of the sunshine and gentle rain that nourish the field simply because science has found them to involve the natural processes of nuclear fusion and gaseous condensation. The matter hinges, rather, on issues of religious dogma, and Christians of good faith can and do disagree. Thus, we are free to adopt this view if we find it rings of truth to us.

Alternatively, one might simply accept that the scientific account of human history is distinct from spiritual accounts in its content and context and not require it to be consistent with one's spiritual knowledge of human origins and human nature (Clouser 2001; Ruse 2001, 2002): "When viewed from the standpoint of this religious focus, then, there is no excuse for treating Genesis as a source of scientific information. It is not good science or bad science because it is not science at all. It is not concerned with how old the earth is, when and how life forms first appeared, or what role the processes of created nature may have had in bringing them about. It is always about God's purposes and chiefly his purposes concerning how humans can stand in proper relation to himself" (Clouser 2001, p. 518). On other matters people routinely make this distinction between spiritual and empirical values. For example, Buddhists accept that meat is potentially a good source of high-quality nutrition, but for spiritual reasons believe it immoral (and therefore an impediment to attaining enlightenment) to eat other sentient beings. We need not deny the nutritional value of meat; the fact is simply irrelevant to the spiritual issues of concern.

One can imagine a number of other paths to resolution as well. But I think my point is clear. We must be careful to approach the world with humility. In particular, we need to be careful about whether or not there is only one reasonable way to understand the world. I think the answer is pretty clear that there isn't one and only one way. There are many approaches and they differ perhaps in degrees of usefulness and insight, degrees of objectivity, but there is no single correct way that works for every issue of human concern or in every context. To argue otherwise is childish at best and at worst is hateful and self-destructive. We will be at our best when we are able to embrace the many complementary ways that we humans come to understand our world and find meaning in our lives.

NOTES

1. By "approach to knowledge" I refer to the basic set of propositions that we hold about the nature of phenomena that we find to be of interest and the consequent ways we go about gaining reliable and useful understandings of these phenomena—how we discover the things we believe to be true and why we accept them as such.

2. I will define what I mean by purpose more fully in a later section, but in the current context what I mean is that the world (including the apt features that organisms exhibit) is as God intended it to be.

3. That is to say, by establishing a natural process, such as natural selection, and then allowing it to run its course, unsupervised.

4. The religious approach described here is dominantly led by Protestant Christians in the United States, but other fundamentalist religious groups follow a very similar path. For instance, a Muslim group based in Istanbul, Turkey, which calls itself the Science Research Foundation, operates an elaborate Web site (www.harunyahya.com) that borrows heavily from U.S. creationists. In recent years their English-language spokesperson, Dr. Oktar Babuna (a medical doctor), has traveled widely in an effort to spread their views.

5. By spirituality, I mean experiences and concerns that focus on the relationship between the self and some sense of the divine. In the Judeo-Christian and Islamic traditions this involves communal and contemplative experiences intended to place oneself in proper relation to God and God's intentions for us. In nontheist traditions such as Buddhism, the concern is to connect through contemplative means to one's true Buddha nature and to cultivate an awareness of the Buddha nature in others. For atheists, a comparable form of spirituality may be the tradition of communing with nature, in which we celebrate organic diversity and wondrous adaptation itself as a manifestation of the divine and seek to position ourselves in an appropriately respectful relation to this meaning of divinity.

REFERENCES

Bambach, R. K., A. H. Knoll, and S. C. Wang. 2004. Origination, Extinction, and Mass Depletions of Marine Diversity. *Paleobiology* 30(4): 522–42.

Behe, M. J. 1996. *Darwin's Black Box: The Biochemical Challenge to Evolution*. Free Press, New York.

Bowler, P. J. 1984. *Evolution: The History of an Idea*. University of California Press, Berkeley.

Brauer, M. J., and D. R. Brumbaugh. 2001. Biology Remystified: The Scientific Claims of the New Creationists. Pp. 289–334 in R. T. Pennock, ed., *Intelligent Design Creationism and Its Critics: Philosophical, Theological, and Scientific Perspectives*. MIT Press, Cambridge, MA.

Chen, X., M. J. Melchin, H. D. Sheets, C. E. Mitchell, and F. Jun-xuan. 2005. Patterns and Processes of Latest Ordovician Graptolite Extinction and Recovery Based on Data from South China. *Journal of Paleontology* 79:842–61.

Clouser, R. 2001. Is Theism Compatible with Evolution? Pp. 513–36 in R. T. Pennock, ed., *Intelligent Design Creationism and Its Critics: Philosophical, Theological, and Scientific Perspectives*. MIT Press, Cambridge, MA.

Crampton, J. S., M. Foote, A. G. Beu, P. A. Maxwell, R. A. Cooper, I. Matcham, B. A. Marshall, and C. M. Jones. 2006. The Ark Was Full! Constant to Declining Cenozoic Shallow Marine Biodiversity on an Isolated Midlatitude Continent. *Paleobiology* 32(4): 509–32.

Darwin, C. 1859. *On the Origin of Species*. Repr., 1909, with an introduction by C. W. Eliot, ed. P. F. Collier and Son, New York.

Dawkins, R. 1986. *The Blind Watchmaker*. Norton, New York.

———. 2006. *The God Delusion*. Houghton Mifflin, Boston, MA.

Dembski, W. A. 1998. *The Design Inference: Eliminating Chance through Small Probabilities*. Cambridge University Press, Cambridge, UK.

Dennett, D. C. 1995. *Darwin's Dangerous Idea: Evolution and the Meanings of Life*. Simon and Schuster, New York.

Dobzhansky, T. 1962. *Mankind Evolving*. Columbia University Press, New York.

Fisher, R. A. 1930. *The Genetical Theory of Natural Selection*. Repr., 1958, rev. ed. Dover, New York.

Gould, S. J. 1988. On Replacing the Idea of Progress with an Operational Notion of Directionality. Pp. 319–38 in M. H. Nitecki, ed., *Evolutionary Progress*. Chicago University Press, Chicago, IL.

———. 1996. *Full House: The Spread of Excellence from Plato to Darwin*. Harmony, New York.

———. 1997. Non-Overlapping Magisteria. *Natural History* 106:16–22. www.stephenjaygould.org/library/gould_noma.html.

———. 1999. *Rocks of Ages: Science and Religion in the Fullness of Life*. Ballantine, New York.

Huxley, J. S. 1942. *Evolution: The Modern Synthesis*, 2nd ed. Allen and Unwin, London.

John Paul II. 1996. Message to Pontifical Academy of Sciences, October 22. *L'Osservatore Romano* 30:3, 7.

Jones, J. E., III. 2005. Memorandum Opinion. *Kitzmiller v. Dover Area School District*. Case 4:04-cv-02688-JEJ. Document 342. U.S. District Court for the Middle District of Pennsylvania, December 20. www.pamd.uscourts.gov/kitzmiller/kitzmiller_342.pdf.

Kitcher, P. 2001. Born-Again Creationism. Pp. 257–88 in R. T. Pennock, ed., *Intelligent Design Creationism and Its Critics: Philosophical, Theological, and Scientific Perspectives*. MIT Press, Cambridge, MA.

Kowalewski, M., W. Kiessling, M. Aberhan, F. T. Fursich, D. Scarponi, S. L. B. Wood, and A. P. Hoffmeister. 2006. Ecological, Taxonomic, and Taphonomic Components of the Post-Paleozoic Increase in Sample-Level Species Diversity of Marine Benthos. *Paleobiology* 32(4): 533–61.

McShea, D. W. 1994. Mechanisms of Large-Scale Evolutionary Trends. *Evolution* 48:1747–63.

Moore, A. L. 1890. The Christian Doctrine of God. Pp. 41–81 in C. Gore, ed., *Lux Mundi: A Series of Studies in the Religion of the Incarnation*, 10th ed. John Murray, London.

Nitecki, M. H., ed. 1988. *Evolutionary Progress*. Chicago University Press, Chicago, IL.

Paley, W. 1809. *Natural Theology; or, Evidences of the Existence and Attributes of the Deity*, 12th ed. J. Faulder, London. (Orig. pub. 1802.)

Peacocke, A. 2001. Welcoming the "Disguised Friend": Darwinism and Divinity. Pp. 471–86 in R. T. Pennock, ed., *Intelligent Design Creationism and Its Critics: Philosophical, Theological, and Scientific Perspectives*. MIT Press, Cambridge, MA.

Pennock, R. T. 1999. *Tower of Babel: The Evidence against the New Creationism*. MIT Press, Cambridge, MA.

———, ed. 2001. *Intelligent Design Creationism and Its Critics: Philosophical, Theological, and Scientific Perspectives*. MIT Press, Cambridge, MA.

Plantinga, A. 1991. When Faith and Reason Clash: Evolution and the Bible. *Christian Scholar's Review* 21(1): 8–32.

Provine, W. B. 1988. Progress in Evolution and the Meaning of Life. Pp. 49–74 in M. H. Nitecki, ed., *Evolutionary Progress*. University of Chicago Press, Chicago, IL.

Raup, D. M. 1988. Testing the Fossil Record for Evolutionary Progress. Pp. 293–318 in M. H. Nitecki, ed., *Evolutionary Progress*. Chicago University Press, Chicago, IL.

Rigby, S. 2003. The Functional Morphology and Population Structure of *Rastrites maximus* from the Southern Uplands. *Scottish Journal of Geology* 39(1): 51–60.

Rigby, S., and B. Rickards. 1989. New Evidence for the Life Habit of Graptolites from Physical Modeling. *Paleobiology* 15(4): 402–13.

Rosslenbroich, B. 2006. The Notion of Progress in Evolutionary Biology: The Unresolved Problem and an Empirical Suggestion. *Biology and Philosophy* 21:41–70.

Ruse, M. 1988. Molecules to Men: Evolutionary Biology and Thoughts on Progress. Pp. 97–126 in M. H. Nitecki, ed., *Evolutionary Progress*. Chicago University Press, Chicago, IL.

———. 1996. *Monad to Man: The Concept of Progress in Evolutionary Biology*. Harvard University Press, Cambridge, MA.

———. 2001. *Can a Darwinian Be a Christian? The Relationship between Science and Religion*. Cambridge University Press, Cambridge, UK.

———. 2002. *The Evolution Wars: A Guide to the Debates*, with a foreword by Edward O. Wilson. Rutgers University Press, New Brunswick, NJ.

Simpson, G. G. 1952. *The Meaning of Evolution: A Study of the History of Life and Its Significance for Man*. Yale University Press, New Haven, CT.

Steinbeck, J. 1945. *Cannery Row*. Repr., 1994, with an introduction by Susan Shillinglaw. Penguin Books, New York.

Van Till, H. J. 2001. The Creation: Intelligently Designed or Optimally Equipped? Pp. 487–512 in R. T. Pennock, ed., *Intelligent Design Creationism and Its Critics: Philosophical, Theological, and Scientific Perspectives*. MIT Press, Cambridge, MA.

Wagner, P. J., M. Kosnik, and S. Lidgard. 2006. Abundance Distributions Imply Elevated Complexity of Post-Paleozoic Marine Ecosystems. *Science* 314:1289–92.

Williams, G. C. 1966. *Adaptation and Natural Selection*. Princeton University Press, Princeton, NJ.

SEVEN • The Misguided Attack on Methodological Naturalism

KEITH B. MILLER

INTRODUCTION

Recent efforts by antievolutionary advocates have focused not so much on science content but on changing the definition of science itself. These efforts are expressions of widely held misunderstandings of the nature and limitations of science. Science is a methodology that provides a limited, but very fruitful, way of knowing about the natural world. This method works only if science confines itself to the investigation of *natural* entities and forces. Scientists seek to understand observations of the natural world only in terms of natural cause-and-effect processes. This self-limitation is sometimes referred to as methodological naturalism. It is the basis for the testability of scientific propositions.

Both traditional creationists and intelligent design (ID) advocates argue that the methodological limitation of science to the study of natural agents and processes is equivalent to the denial of the existence and action of God. This is a reflection of their false claim that science as currently practiced is inherently atheistic. It is a fundamental confusion of methodological naturalism with philosophical naturalism or materialism. Philosophical naturalism is the belief that the material universe is all that there is and that scientific knowledge exhausts all potential knowledge about what is real. It is a philosophical claim, not a statement about the nature of science.

The proponents of intelligent design are also committed to the belief that God's action is scientifically detectable—that divine action is subject to scientific inquiry.

Accordingly, they argue that science must include the action of intelligent supernatural agents. They see methodological naturalism not as a description of the limitations of scientific inquiry, but as an arbitrary and unjustified prescription that prevents scientists from including supernatural action in their scientific explanations. However, these attacks on methodological naturalism are misguided and reveal a failure to distinguish between natural and supernatural agency, and to recognize that scientific descriptions, however complete, pose no threat to theological understandings of the action of God in nature.

Methodological naturalism has thus become the focal point of intense criticism by both traditional creationists and intelligent design advocates. Widespread efforts have been made to challenge the methodology of science on grounds that it promotes an atheistic worldview, undermines moral values, and discriminates against religious faith.

THE ATTACK AGAINST METHODOLOGICAL NATURALISM

Probably one of the best case studies illustrating how the nature of science has come under attack in recent years is the widely reported effort of young-Earth creationists and intelligent design advocates in Kansas to rewrite the state science standards. These efforts, begun in 1999 and still going on, have centered on the nature of science, and particularly on an attack on methodological naturalism as a description of the scientific process.

The single word *natural* is the focus of considerable attention. In the science standards recommended in 1999 by an appointed committee of scientists and science educators and teachers (and subsequently rejected by a majority of the Kansas State Board of Education), science is appropriately defined as follows: "Science is the human activity of seeking natural explanations for what we observe in the world around us."[1] Although this may seem a simple, uncontroversial statement, the creationists and ID supporters saw in this wording an attempt to promote an atheistic and materialistic worldview. They removed the word *natural* from this definition of science, and replaced it with the word *logical.* This action was not just minor wordsmithing. The purpose of the change was to enable inclusion of nonnatural or supernatural explanations and agents as part of the scientific enterprise.

The rationale for the change was explained by John Calvert, director of the Intelligent Design Network and one of the primary advocates for the standards changes: "The word 'natural' is a code word for the non-consideration of any teleological or purposeful explanation of the universe. . . . This notion of exclusion of teleological data from consideration has become an unwritten law within the scientific community. It has become the major tenet of a faith that everything, including our lives, can be explained only by material things without resort to an intelligent agent."[2] Note how this claim distorts the simple meaning of the earlier definition of science to imply both a deliberate exclusion of relevant scientific data and the promotion of a materialistic "faith" that denies purpose and meaning in our lives. This statement by Calvert also implies that science has no methodological boundaries. The replacement of the word *natural* with *logical* so broadens the definition of *science* as to make it virtually synonymous with *knowledge*.

The national science standards have come under the same attack, as seen in the following words of Jody Sjogren, another Intelligent Design Network spokesperson: "By contrast, the National Standards would limit teaching to only 'natural explanations,' so that a teacher could teach only one side of the controversy about the cause of life and its diversity. The evidence supporting design would be ignored, not because it doesn't exist, but because of an a priori philosophical assumption that natural causes are all there is."[3] As in the previous statement by Calvert, Sjogren here portrays science as being unjustifiably exclusive and based on a philosophy that denies the existence of God. This quote is also clear in expressing the view that design arguments will be considered as part of scientific explanation only if science is defined to include nonnatural (i.e., supernatural) causes. This view explains the intense focus by ID advocates on getting their definition of science accepted in the state science standards.

In their more recent challenges to Kansas science education, ID advocates have explicitly attacked methodological naturalism as a description of the nature and limitations of science. They justified their rewrite of the definition of science in the standards as follows: "The current definition of science is intended to reflect a concept called methodological naturalism, which irrefutably assumes that cause-and-effect laws (as of physics and chemistry) are adequate to account for all phenomena and that teleological or design conceptions of nature are invalid. Although called a 'method of science,' the effect of its use is to limit inquiry (and permissible explanations) and thus to promote the philosophy of Naturalism."[4] Methodological naturalism is here deliberately conflated with philosophical naturalism. The central objective of the creationists and ID advocates is to so firmly associate the methods of modern science with a materialistic and atheistic worldview that supernatural explanations in the form of ID arguments can be admitted as a "balance."

The central role of the redefinition of science in the arguments of ID proponents was evident in the 2005 intelligent design case in Dover, Pennsylvania. During that trial, testimony by ID advocates and references to the ID literature demonstrated that the acceptance of ID as a scientific argument depended on the rejection of methodological naturalism as a description of science and the inclusion of the supernatural. In his opinion, Judge John E. Jones III concluded that ID is fundamentally a religious and philosophical argument, and not a scientific alternative to evolution. He further concluded that although scientific theories do not appeal to the existence or action of God, evolution "in no way conflicts with, nor does it deny, the existence of a divine creator."[5]

The reaction of John Calvert to the decision by Judge Jones further illustrates the degree to which most ID supporters firmly believe that the methods of modern science, and evolution in particular, promote an atheistic worldview: "Evolution, and the naturalism which effectively shields it from scientific criticism, is key to all of the major non-theistic religions and belief systems. The Dover opinion censors scientific data that is friendly to one set of religious beliefs in favor of data that supports competing and antagonistic belief systems. For the Court, it is OK for the state to put into the minds of impressionable students evidence that promotes a materialistic and non-theistic world view while censoring contradictory evidence that supports a theistic one."[6]

As suggested by the Kansas and Dover cases, the rejection of methodological naturalism (MN) and its conflation with philosophical naturalism is not an isolated phenomenon but rather a central element of the arguments of ID advocates. This is made clear in the writings of Phillip Johnson—recognized by many as a principal founder of the ID movement: "We [members of the intelligent design movement] are opposed by persons who endorse methodological naturalism, a doctrine that insists that science must explain biological creation only by natural processes, meaning unintelligent processes. Reference to a creator or designer is relegated to the realm of religion, and ruled out of bounds in science regardless of the evidence."[7] Note that MN is treated by ID advocates as a doctrine, a philosophical assumption, rather than a methodological limitation of scientific inquiry. That is, they see the practice of science as based on a philosophy that claims that the material universe is all that exists. Furthermore, MN is viewed as excluding real scientific evidence from the discussion and restricting the search for truth. Johnson's conflation of MN and philosophical naturalism is made explicit in this following statement from his influential book *Darwin on Trial:*

Make no mistake about it, in the Darwinist view, which is the official view of mainstream science, God had nothing to do with evolution. Theistic or "guided" evolution has to be excluded as a possibility because Darwinists identify science with a philosophical doctrine known as *naturalism*. Naturalism assumes the entire realm of nature to be a closed system of material causes and effects, which cannot be influenced by anything from "outside." Naturalism does not explicitly deny the mere existence of God, but it does deny that a supernatural being could in any way influence natural events, such as evolution, or communicate with natural creatures like ourselves. *Scientific* naturalism makes the same point by starting with the assumption that science, which studies only the natural, is our only reliable path to knowledge. A God who can never do anything that makes a difference, and of whom we can have no reliable knowledge, is of no importance to us.[8]

Johnson clearly perceives science to be a thinly disguised effort to promote a godless worldview. Evolution in particular is seen as inherently atheistic and inseparably wedded to a worldview that denies, if not God's existence, then at least God's involvement in the natural world. Also, as seen in the earlier example, evolution is consistently referred to by the intended pejorative term *Darwinism* by nearly all ID advocates. This effectively associates evolutionary theory with a term that has already been invested in the public mind with negative social and political ideologies.

Traditional creationists and most intelligent design advocates believe fundamentally that evolutionary theory and orthodox Christian faith are in irreconcilable conflict with each other. This conviction is also a central force behind the political strategy of the ID movement. Phillip Johnson sees the objective of the ID movement as framing the public debate over evolution in terms of atheism versus theism.[9] Thus, for many in the ID movement, the conflict is not between scientific theories, or between different views of the place of the supernatural in science; the conflict is a battle between theism and atheism.

The attack on methodological naturalism and its conflation with philosophical naturalism is not a new development of the intelligent design movement—instead it is a reframing of the same false dichotomy that undergirds the arguments of "creation science." This view is clearly stated by two of the founders of creation science, Henry Morris and Gary Parker:

It [the creation/evolution question] deals with two opposing basic world views—two philosophies of origins and destinies, of life and meaning. . . .

> One of these two world views—evolution—assumes that the universe is self-contained, and that the origin and development of all its complex systems (the universe, living organisms, man, etc.) can be explained solely by time, chance, and continuing natural processes, innate in the very structure of matter and energy.
>
> The second world view—creation—maintains that the universe is *not* self-contained, but that it must have been created by processes which are not continuing as natural processes in the present. One or the other of these two philosophies (or "models," as they are frequently called) must be true, since there are only these two possibilities. That is, all things either can—or cannot—be explained in terms of a self-contained universe by ongoing natural processes. . . . The Evolution Model, by its very nature, is an atheistic model (even though not all evolutionists are atheists) since it purports to explain everything without God. The Creation Model, by *its* nature, is a theistic model (even though not all creationists believe in a personal God) since it requires a creator able to create the whole cosmos.[10]

The similarity between this statement and the preceding one by Phillip Johnson is quite striking. Both set up a false dichotomy of two mutually exclusive and exhaustive alternatives. Modern evolutionary science is portrayed as based on a fully materialistic and atheistic worldview, and only intelligent design or creationism is provided as an alternative. Interestingly, the parsing of all causal explanations into those of "time," "chance," "natural processes," and "creation," by Morris and Parker, is also a close parallel to the use of the categories "chance," "natural regularities," and "design" in the "explanatory filter" of William Dembski, a prominent ID advocate.[11]

The rejection of MN and the broadening of science to include all forms of human knowledge and inquiry is also explicit in creation science writings, such as that by Morris and Parker: "But who ever defined 'science' as 'naturalism'? The word *science*, comes from the Latin *scientia*, meaning 'knowledge.' To assume that knowledge can be acquired solely on the assumption of naturalism is to beg the question altogether. Scientists are supposed to 'search for truth,' wherever that search leads."[12] This is precisely the view of science that the ID advocates tried to advance by removing the word *natural* from the definition of science in the Kansas state science standards. Because supernatural agents are excluded from scientific description, ID advocates see science as a tool of an atheistic worldview intent on removing all mention of God from the culture.

The dichotomizing views of both ID and creation science advocates are an extension of the widely held warfare view of science and faith. That warfare metaphor owes much of its modern expression to a pair of widely influential nineteenth-century works: John William Draper's *History of the Conflict between Religion and Science* (1874) and Andrew Dickson White's *A History of the Warfare of Science with Theology in Christendom* (1896). Such views were based on, and have been perpetuated by, simplistic and grossly inaccurate readings of history. This warfare view has since been discredited by both theological and historical scholarship.[13] Most Christian theologians, including evangelicals, have long recognized that a faithful reading of the Bible does not demand a young Earth, nor does it prohibit God's use of evolutionary mechanisms to accomplish God's creative will. To the present day, Christian scientists and theologians have articulated this integration of evolutionary science and Christian faith within a broad range of theological traditions.[14]

THE ORIGIN AND MEANING OF METHODOLOGICAL NATURALISM

After a review of the charges and claims made against methodological naturalism as a description of modern science, it is appropriate to look at the meaning and context of the term as it was originally proposed. Possibly the earliest detailed use and discussion of the term was in 1986 by Paul deVries, an evangelical Christian philosopher at Wheaton College. He used the term *methodological naturalism* to describe the legitimate purview of science as one limited to explaining and interpreting the natural world in terms of natural processes and causes. He describes scientific inquiry as follows: "The goal of inquiry in the natural sciences is to establish explanations of contingent natural phenomena strictly in terms of other contingent natural things—laws, fields, probabilities. Any explanations that make reference to supernatural beings or powers are certainly excluded from natural science. . . . The natural sciences are limited by method to naturalistic foci. By method they must seek answers to their questions within nature, within the non-personal and contingent created order, and not anywhere else. Thus, the natural sciences are guided by what I call methodological naturalism."[15] This is precisely the same understanding that was used in the definition of science rejected by ID advocates in Kansas.

Paul deVries embraced this understanding of the nature and limitations of science because he saw it as consistent with, and supportive of, a vibrant and vital role for theology: "If we are free to let the natural sciences be limited to their perspectives under the guidance of methodological naturalism, then other sources of truth will be more defensible. However, to insist that God-talk be included in the natural sciences is to submit unwisely to the modern myth of scientism: the myth that all truth is scientific."[16] He argues that MN gives proper intellectual space to theological inquiry and rejects science as the ultimate arbiter of all truth claims. In his view, to broaden science to include the supernatural would be yielding to a culture of scientism. Thus, the young-Earth creationists and ID advocates have understood MN in exactly the opposite manner from its originally intended meaning. Ironically, by rejecting methodological naturalism, ID advocates have ended up supporting the very scientism that they claim to want to fight against. They are seeking to construct science as an all-encompassing search for truth.

Although deVries thought that the term *methodological naturalism* was original with him,[17] Edgar Brightman used it much earlier in a 1936 presidential address to the eastern division of the American Philosophical Association. The focus of Brightman's paper is "metaphysical naturalism," which he usually just refers to as "naturalism." He introduces methodological naturalism but does not really develop the idea further than to distinguish it from metaphysical naturalism: "Every thinking experient will, in some sense, reach the stage of naturalism. He will accept nature as the space-time order described by the sciences. . . . Such a universal naturalism—common to idealists and realists, to naturalists and theists alike—may be called scientific or methodological naturalism. But methodological naturalism is sharply to be distinguished from metaphysical naturalism. The latter takes the incomplete description and heuristic methods of the former to be either final truth about reality or at least the limits of present human knowledge."[18]

Brightman further argues that theology and philosophy are valid paths to knowledge and that they address aspects of human experience that naturalism cannot—aspects such as mystical experience, purpose and meaning, teleology, and values. His purpose is to lay out his philosophical argument for the existence and study of God. Although Brightman does not develop the meaning and significance of methodological naturalism as fully as deVries does, both see science as a limited enterprise restricted to describing nature in terms of nature, and leaving open other pathways to knowledge. Both of them use MN as a description of science to validate the role of theology. They maintain that the limitations of scientific explanation allow other human searches for meaning and truth an authentic place in intellectual life.

Rather than being a prescriptive rule or doctrine, methodological naturalism simply describes what empirical inquiry is. It was never intended as a statement of the nature of ultimate reality. Science as a discipline is silent on the existence or action of God; nor does science deny the existence of a creator. Science does not, and cannot, assert that only material things exist. Some nontheists, seeing God as an unnecessary addition to a scientific description of the universe, extend this to a philosophical exclusion. Although divine action is irrelevant to scientific description, the existence or action of God cannot be thereby dismissed on scientific grounds. Scientific methodology excludes appeals to supernatural agents simply because the action of such agents cannot be tested. Therefore, the argument that there is no reality beyond matter and energy is a philosophical claim, not a scientific one. Such metaphysical naturalism, to use Brightman's phrase, reflects a prior commitment that is then superimposed on science.

It is similarly an error to argue that acceptance of God's existence requires the incorporation of divine action into scientific description. Science can be neither atheistic nor theistic. The scientific enterprise is religiously neutral in that its conclusions about the natural world can be tested against observations of that world, but not against the dictates of any particular faith.

A very important feature of the scientific enterprise is that it takes place within a multicultural and interfaith community of scholars. At a typical professional scientific meeting there will be participants from a wide range of nationalities, cultures, and religious traditions. Yet those scientists from these various backgrounds can sit down together and productively discuss scientific questions, examine evidence, and reach conclusions. They can do this because scientific knowledge is not tied to a particular religious or nonreligious worldview—it is universally accessible. Any attempt to incorporate supernatural action into scientific description, or conversely to declare that science is inherently atheistic, undermines this religious neutrality.

Though science as a discipline is religiously neutral, individual scientists are not—nor must they be. A significant shortcoming of much public discussion of science and faith is the failure to distinguish between the conclusions of scientific research and the interpretations of those conclusions in light of an individual's worldview. We are all complex intellectual beings, and we may each hold strongly to particular philosophical and religious positions. Individual scientists may integrate the conclusions of science into a broader comprehensive philosophical or theological worldview—but such an integrated view does not itself constitute a scientific conclusion. Furthermore, advances in scientific knowledge and technology raise important philosophical, ethical, and theological issues that cannot be resolved through science alone.

THE ATTEMPT TO DEVELOP A THEISTIC SCIENCE

SCIENCE AND THE SUPERNATURAL

Both traditional creationists and ID supporters seek to have nonnatural or supernatural action accepted as a legitimate part of scientific explanation. They contend that science pursued under theistic assumptions must differ in its scientific conclusions from science as currently practiced. There is a strong desire to see scientific evidence for divine action, to have theistic arguments be part of science.

For ID advocates, the argument for design must be included as an accepted part of scientific explanation. God's action must be scientifically detectable. Phillip Johnson views with disdain theological views that do not require God's action to be empirically testable: "God-guided evolution would be genuinely theistic, but the doctrine of methodological naturalism rules out the possibility that God did the guiding in any way that is testable. . . . The theism is in the mind (or faith) of the believer."[19]

Similarly, William Dembski has stated, "Intelligent design is incompatible with what typically is meant by theistic evolution. Theistic evolution takes the Darwinian picture of the biological world and baptizes it, identifying this picture with the way God created life. When boiled down to its scientific content, however, theistic evolution is no different from atheistic evolution, treating only undirected processes in the origin and development of life. . . . The current theological fashion prefers an evolutionary God inaccessible to scientific scrutiny over a designer God whose actions in nature are clearly detectable."[20]

Note that evolutionary theory is presented as though it could be atheistic or theistic. Furthermore, theistic science must somehow look different from atheistic science. Dembski's objective is to develop a theistic science that incorporates divine action as part of its scientific descriptions—to include "God-talk" as deVries would say. But all of this assumes that divine action is subject to empirical test in the first place—an issue that will be discussed later.

Dembski offers further insight into the reasons for insisting that divine action be made part of scientific discourse:

> It bears repeating: the only universally valid form of knowledge within our culture is science. Within late-twentieth-century Western society neither

> religion, nor philosophy, nor literature, nor music nor art makes any such cognitive claim. Religion in particular is seen as making no universal claims that are obligatory across the board. The contrast with science is stark. Science has given us technology—computers that work as much here as they do in the Third World. Science has cured our diseases. Whether we are black, red, yellow or white, the same antibiotics cure the same infections. It is therefore clear why relegating intelligent design to any realm other than science (e.g. religion) ensures that naturalistic evolution will remain the only intellectually respectable option for the explanation of life.[21]

Dembski argues that ID must be brought into the realm of science, not because ID offers a fruitful theoretical framework for scientific discovery, but because to his mind science has cultural power and religion does not. The effort to have ID accepted as valid science is perceived as part of a cultural and worldview battle in which science is solidly in the materialistic/atheistic camp. However, as we have seen, this science/faith warfare view is based on empty rhetoric, not historical reality. Faith claims are important, and should be part of the public and academic discussion. However, neither religion nor science profits by the two being made antagonists in a cultural war.

Dembski also acknowledges that science is transcultural and does not depend on the worldview of the practicing scientist. The injection of religion into science sought by ID advocates threatens to destroy the cultural neutrality that has enabled scientists to work as part of a world community. Furthermore, the understanding of science promoted by ID advocates is undergirded by a particular theological view—one that is not even held by many Christians. Such a view of science cannot become the basis for a productive cross-cultural scientific enterprise.

ATTEMPTING TO DETECT DIVINE ACTION

The absence of references to supernatural causes in scientific description is not just an agreed-on philosophical limitation of science, but a consequence of the practical inability of science to detect divine action. It is worth noting that even ID advocates employ methodological naturalism in their own scientific research for the simple reason that the actions of nonnatural agents cannot be incorporated into a scientific research program. Typically, ID proponents overlay their philosophical and religious understandings on scientific conclusions. That is, they invest particular scientific observations with theological meaning. Though anyone is free to apply his or her religious and philosophical perspectives to the interpretation of science, such philosophical perspectives are not themselves scientific.

From the perspective of scientific inquiry, a supernatural agent is effectively a black box, and appeals to supernatural action are essentially appeals to ignorance. A supernatural agent is unconstrained by natural laws or the properties and capabilities of natural entities and forces—it can act in any way and accomplish any conceivable end. As a result, appeals to such agents can provide no insight into understanding the mechanisms by which a particular observed or historical event occurred. Belief in the creative action of a supernatural agent does not answer the question of how something happens. "A miracle occurs here" is no more an answer to the question of "How?" than is "We don't know."

This same point can be made beginning from a theological perspective. As understood in Christian theology, divine action includes the doctrine of providence, which concerns God's sustaining and upholding of the natural world, and divine cooperation with, and governance of, nature.[22] Divine action in this sense does not imply any break in the continuity of cause-and-effect natural processes. An internally complete scientific description would be consistent with this theological view. Consequently, scientific and theological understandings can be seen as complementary—science provides a description of natural phenomena as they are upheld by divine providence.

However, what of divine miracle? The traditional Christian theological understanding of miracle is that of a sign that draws attention to God's character or will—it carries theological meaning. A miracle in this sense does not require that the sign break natural law or interrupt natural chains of cause and effect. Only the subset of miraculous actions that involve divine intervention and that break natural chains of cause and effect are potentially in conflict with scientific explanations.

Can law-breaking, miraculous events in natural history be detected or falsified scientifically? Although not falsifiable, a specific claim of divine action of this kind could be brought into question if a series of natural cause-and-effect processes could be shown to plausibly account for the miracle. However, such a conclusion says nothing about God's action in and through those processes. If, on the other hand, no plausible series of natural processes is currently known to account for the miracle, then scientists will continue to search for such a natural explanation. A true break in the continuity of natural processes is indistinguishable from current ignorance.

Scientific investigation cannot conclude that a particular event in the history of life, or a particular feature of the natural world, must be the consequence of a supernatural agent. We are of course free to make those claims from a theological perspective. But those claims must be evaluated on their theological and philosophical merits. Many Christian theists believe that tying our arguments for the creative action of God to failures in scientific description poses theological problems. In conclusion, while theists and nontheists can debate the important questions of the existence of God, and whether God does or does not act outside the regularities of the natural world, science has little or nothing to say about it. The scientific study of the natural world simply cannot confirm or deny the existence or action of God.

THE LIMITATIONS OF SCIENCE AND THE SEARCH FOR "TRUTH"

ID advocates claim that the exclusion of God from scientific description unnecessarily restricts the search for truth. They argue that the limitation of science to describing nature solely in terms of natural forces and processes results in erroneous conclusions, and in a failure to discover the truth about the natural world. Paul Nelson, a prominent ID advocate and spokesperson, and John Marks Reynolds have made this point as follows:

> Christians who are theistic evolutionists are in a cruel bind. Their theology demands a God who acts in space and time. They are captured, however, by a methodological naturalism in science that will not allow them to scientifically consider positive evidence for a creator. They are so fearful of being wrong about proclaiming God's activity in the natural world that they have decided that his activity is invisible to human science. As we shall see, this limitation of science impedes the ability of theistic evolutionists to consider all the possibilities. . . . If God is at the bottom of it all, and only a naturalistic answer is acceptable to science, then in the end natural science will be left with a gap in its knowledge. Without an ability to turn to the supernatural, science will be left with hopelessly false naturalistic speculations about the reason physical objects exist. Worse still, a prior commitment to methodological naturalism will lead these selfsame scientists to continue looking for a naturalistic basis for existence when there is none. The entire position is doomed to research futility.[23]

William Dembski has argued similarly: "Is there a code of scientific correctness that instead of helping lead us into truth, actively prevents us from asking certain questions and thereby coming to the truth? We are dealing here with something more than a straightforward determination of scientific facts or confirmation of scientific theories. Rather we are dealing with competing worldviews and incompatible metaphysical systems. In the creation-evolution controversy we are dealing with a naturalistic metaphysic that shapes and controls what theories of biological origins are permitted on the playing field in advance of any discussion or weighing of evidence."[24]

These passages confuse scientific description with a statement of the nature of all reality—they conflate methodology and worldview. They make the fundamental error of viewing science as equivalent to knowledge and ordaining it the arbiter of all truth. As mentioned earlier, science cannot test divine action empirically. Unless it is assumed that science is the only source of truth, then acknowledging the limitations of scientific investigation does not curtail the pursuit of that truth.

Contrary to the ID claim, if God is assumed to be active in the natural world, then the methodological naturalism of science does not necessitate "hopelessly false naturalistic speculations" about the natural world. If we assume, for the sake of argument, that God acted in creation to bring about an event in a way that broke causal chains, then science will conclude that no series of cause-and-effect processes are presently known that can adequately account for this event, and research will continue to search for such processes. Any statement beyond that would require the application of a particular religious worldview. Science could not conclude that "God did it." Given our incomplete knowledge, we could not be sure an exhaustive search for cause-and-effect processes was ever complete. Alternatively, if God acted through a seamless series of cause-and-effect processes, then, stimulated by the tentativeness and methodological naturalism of science, scientists might discover those processes. By contrast, using an ID approach, an investigator would make the inference to intelligent design, and any motivation for further research would end. Thus, ID runs the risk of drawing false conclusions, and prematurely terminating the search for cause-and-effect descriptions when one isn't already at hand. Gaps in our knowledge can be filled only by continued efforts to search for possible natural causes through research conducted using the assumption of methodological naturalism.[25]

The risk of prematurely ending the search for cause-and-effect explanations is a real one if the perspective of ID advocates is adopted. In fact, they see much of current frontier research as futile. Nelson states, "Skeptics of design ask design theorists to go fishing for causes where there is no reason to think the fish will be caught: certainly the skeptics have caught none. The skeptic of design, a philosophical or methodological naturalist, typically, asks us to pursue the naturalistic program of explanation without reason."[26] How does one determine that the search for a cause-and-effect answer is without reason? Appealing to current knowledge will not do, unless one claims complete knowledge of all natural processes.

Such a view strikes at the heart of the scientific enterprise. Science is driven by the desire to resolve currently unsolved problems concerning Earth and cosmic history and the mechanisms of the natural world. The open questions about our natural world motivate the work of the scientific community and inspire the work of individual scientists. The goal of science is not merely to describe the known, but to press the boundaries of the unknown. Research in chemical evolution and the origins of life, for example, is still far from a consensus view on likely pathways to the first self-replicating life-form. However, it has enlarged our understanding of some fundamental aspects of biological systems in extreme environments, provided new insights into abiogenic organic synthesis, improved our understanding of early Earth environments, and so on. I see no reason why that research will not continue to be fruitful and continue to resolve outstanding problems.

Oddly, some ID advocates have already declared research into the origin of life to be a futile and unreasoned pursuit. Phillip Johnson, for example, declares that research on the origin of life has come to a dead end: "Why not consider the possibility that life is what it so evidently seems to be, the product of creative intelligence? Science would not come to an end, because the task would remain of deciphering the languages in which genetic information is communicated, and in general finding out how the whole system *works*. What scientists would lose is not an inspiring research program, but the illusion of total mastery of nature."[27] But a substantial field of science would end, for the ID view implies that solutions to physical, chemical, and biological questions concerning the origin and early evolution of life are not attainable. The logical consequence of such a perspective is that all such research should cease, since it constitutes a waste of resources and an exercise in futility.

Though traditional creationists and ID advocates both talk about pursuing the evidence wherever it leads, they do not practice it. With regard to the scientific evidence, young-Earth creationists Nelson and Reynolds state, "Natural science at the moment seems to overwhelmingly point to an old cosmos. Though creationist scientists have suggested some evidences for a recent cosmos, none are widely accepted as true. It is safe to say that most recent creationists are motivated by religious concerns."[28]

If the goal is to follow the evidence where it leads, then surely they would readily accept this overwhelming evidence for an ancient Earth and cosmos. Yet, they remain young-Earth creationists, defending their position as follows: "There are, however, two very good reasons to maintain a young earth position during the struggle. First, recent creationism is intellectually interesting. . . . Second, a coherent recent creationism would be a great boon to religious belief."[29] Given arguments such as these, it seems disingenuous for ID advocates and young-Earth creationists to criticize the scientific community for placing metaphysics ahead of evidence.

METHODOLOGICAL NATURALISM, "INTELLIGENCE," AND THE DESIGN FILTER

IS "INTELLIGENCE" NATURAL?

The "design filter" is one of the primary ID arguments. It is based on a parsing of the causes of all phenomena into three categories that are both nonoverlapping and exhaustive: chance, natural regularities (or natural law), and design.[30] This scheme was developed by William Dembski as a means of detecting design, although, as we have seen, it has parallels in the earlier creation science literature. The fundamental argument is that the elimination of chance and natural law as plausible causes for an event or structure leaves design as a default answer. Significantly, ID advocates make no explicit distinction between natural and supernatural action. Design becomes identified with nonnatural or "intelligent" causes. This line of reasoning eliminates intelligence from the realm of the natural and places it in the same category as the supernatural. Nelson makes clear this dichotomy between *natural* and *intelligent* cause: "The explanatory filter (Dembski 1998), however, reliably discriminates between naturally and intelligently caused phenomena. It does so because naturally caused phenomena are trapped by their corresponding causes or mechanisms, necessarily, as a matter of method, at the first and second analytical nodes of the filter. Thus any object or pattern for which we have a sufficient natural cause cannot be assigned to design."[31]

ID advocates reject humans as natural agents, and instead view them as nonnatural, intelligent agents distinct from the natural world. Human (and humanlike) agents and supernatural agents are viewed as essentially identical categories with respect to scientific explanation. Thus a demonstration of human intelligent action is for them indistinguishable from a demonstration of divine action. This equation of human and divine action is crucial for their argument that supernatural intelligence can be detected empirically.

Nelson argues for the nonnatural character of human actions:

> At this very moment, you are engaged in a nonnaturalistic event. Traditional Christianity teaches that your nonphysical soul is engaged with your body in the task of reading. You are, if science must be naturalistic, engaged in an activity that science will never understand. Science bound by naturalism will never be able to recognize an immaterial soul. Reading is not scientifically explainable. This holds true for whatever activity in which humans, or any other beings with souls, engage themselves. Worst of all, the same research futility that plagued the physicist will return with a vengeance for the psychologist. Human psychology, if it can only recognize natural causes for events, will be forever on the hapless task of trying to explain the actions of the soul without including the soul in the theory.[32]

Nelson's logic is incorrect. However one understands human soulishness, humans are *natural* causal agents. *Intelligence* is not a distinct category from *natural.* Humans are part of nature—in fact a part of nature about which we know a considerable amount. Human behaviors and physical capabilities are known, and we can therefore recognize the artifacts produced by them. Because of this knowledge, past human actions and behavior can be reconstructed from archaeological or fossil records. The detection of past human actions is also not different in principle from that of the past purposeful action and behavior of long extinct animals. Paleontologists can, for example, study the patterns of breakage on shells or bones to infer the behavior and identity of a likely predator. Even animal burrows can be recognized and distinguished from chance markings and attributed to specific behaviors such as feeding or escape. We can infer much about the behavior and interactions of organisms from the fossil record. All such purposeful actions are part of the natural world that is subject to scientific study.

By contrast with natural agents, supernatural agents are not part of nature and thus not subject to empirical test. As we have already seen, the actions of supernatural agents are unconstrained by physical law and thus are effectively equivalent to statements of ignorance. It is invalid to argue that our ability to recognize products of past human actions is equivalent to identifying supernatural action. The supernatural is not just the natural writ large—it is *other* than nature. God is not Superman.

Thus, there are at least three major arguments against the design filter as a means to detect the action of supernatural agents: (1) humans are intelligent agents that are also natural agents, and cannot be used as models of the supernatural; (2) design, when used as a default explanation in the absence of a currently satisfactory natural explanation, is indistinguishable from ignorance; and (3) contrary to the claims of ID proponents, actions by natural intelligences are not excluded from scientific examination, as is acknowledged in their own frequent references to archaeology, forensics, and the search for extraterrestrial intelligence (SETI). Such work is fully consistent with the methodological naturalism of science.

DOES METHODOLOGICAL NATURALISM EXCLUDE DESIGN?

In common parlance, the word *design* is used in several different ways: (1) to establish a purpose or goal; (2) to thoughtfully conceive a plan to achieve the goal; or (3) to implement the plan or assemble the structure. The manner of assembly is not restricted or predetermined and is often considered separate from design.

As used by ID advocates, *design* seems to be vaguely and inconsistently defined. We have already seen that *designed* is often used synonymously with *intelligently caused* and *nonnatural* or *supernatural*. In his book *Darwin's Black Box*, ID proponent Michael Behe provides this concise definition: "What is 'design'? Design is simply the purposeful arrangement of parts."[33] This definition seems to combine the ideas of both thoughtful conception and assembly. Such an understanding is reflected in another statement by Behe in that same passage: "To a person who does not feel obliged to restrict his search to unintelligent causes, the straightforward conclusion is that many biochemical systems were designed. They were designed not by the laws of nature, not by chance and necessity; rather, they were *planned*. The designer knew what the systems would look like when they were completed, then took steps to bring the systems about" (author's emphasis).[34]

Note that Behe refers to the design filter—with design set against the categories of chance and necessity (laws of nature). The designer here clearly refers to a divine agent. However, the assembly or fabrication of a natural object or system by a divine agent need not involve any nonnatural process: the implementation of design may be in and through natural processes. How can Behe know a priori that the designer did not use the "laws of nature" and "chance" to accomplish its creative will? Such setting up of design as an alternative to natural process creates a false dichotomy.

There is no inherent conflict between divine design, in the sense of divine purpose and thoughtful conception, and the efforts of science to uncover natural cause-and-effect processes. Divine design in this sense is based on theological argument, not the scientific study of nature.[35] Similarly, the implementation of divine will through the built-in capacities of nature poses no tension with scientific description. However, when divine design is identified as the miraculous (law-breaking) assembly of natural systems and objects, as it seems to be by ID advocates, then it becomes equivalent to a gap in our current scientific knowledge. Such gaps will simply be a current scientific mystery, and that mystery may eventually yield to a natural cause-and-effect explanation.

It is central to any coherent understanding of design that the purposes and capacities of the designer be known. However, ID advocates such as Behe argue that design can be recognized in the absence of any knowledge of the designer: "The conclusion that something was designed can be made quite independently of knowledge of the designer. As a matter of procedure, the design must first be apprehended before there can be any further question about the designer. The inference to design can be held with all the firmness that is possible in this world, without knowing anything about the designer."[36] But the claim is clearly false. We must have some conception of the capabilities (and limitations) of potential causal agents before they can be invoked. We do in fact know much about human designers as a class of potential agents, even if we do not know the specific individuals. We recognize human artifacts because we understand human capacities and purposes. Similarly, we recognize the products of other natural volitional agents, such as nonhuman animals. We can search for the signals of extraterrestrials, but only to the extent that we assume some specific capabilities and purposes on their part (usually modeled after our own). Divine agents, on the other hand, have no constraints, and their purposes and capabilities cannot be defined without reference to a particular theology. We do not know how a divine agent might work in nature. To borrow a phrase from Behe, such agents are "black boxes" that can explain any observation—and thus have no scientific utility.

THE NATURE OF SCIENCE, SCIENCE LITERACY, AND PUBLIC EDUCATION

Popular misunderstandings of the nature and practice of science are fundamental and present a great obstacle to scientific literacy. The rejection of methodological naturalism by ID advocates and traditional creationists supports and perpetuates these misunderstandings. As we have seen, those who oppose the current conclusions of the historical sciences commonly see scientific and theological descriptions of reality as being mutually exclusive and contradictory. These opponents adopt a warfare view of science and faith.

The perception by many that science, particularly geology and evolutionary biology, is a thinly disguised effort to promote a godless worldview is usually accompanied by other false understandings. Theories are viewed as unsubstantiated guesses rather than the unifying concepts that make sense of observations. Theories in the historical sciences are seen as inherently untestable and outside the realm of "true science." Science is understood as the encyclopedic accumulation of unchanging "facts," rather than as a dynamic and human intellectual enterprise. Unless these foundational issues are addressed, further progress toward the public understanding of science will not be forthcoming.[37]

Young-Earth creationist and ID efforts to influence public science education have advanced by associating consensus science with materialism and atheism in the public mind. Equating methodological naturalism with philosophical naturalism is essential to that effort. Once science is perceived as supporting a godless worldview, then ID can readily be admitted as a counterbalance. But, ID must be recognized as a valid scientific alternative in order to pass legal challenges in the public classroom. And once again, the attack on methodological naturalism (MN) is central.

However, the arguments of ID proponents against MN fail repeatedly. MN is not an enforced doctrine or prescription, but a description of what types of questions science can and cannot address. ID advocates themselves have conducted no research that validates the rejection of MN. ID has provided no novel alternative explanations or research program, and has thus failed to contribute anything to science. The ID attack on MN is an effort to gain admittance of their arguments into public science classrooms by definitional fiat, in the absence of productive ID scientific theories or research.

Several ID advocates have admitted to the absence of any significant ID theory or research program, as Paul Nelson stated in 2004: "Easily the biggest challenge facing the ID community is to develop a full-fledged theory of biological design. We don't have such a theory right now, and that's a real problem. Without a theory, it's very hard to know where to direct your research focus. Right now, we've got a bag of powerful intuitions, and a handful of notions such as 'irreducible complexity' and 'specified complexity'—but, as yet, no general theory of biological design."[38] Similarly, Dembski stated in an address delivered in 2002: "Intelligent design as a scientific and intellectual project has many sympathizers but few workers. The scholarly side of our movement at this time consists of a handful of academics and independent researchers."[39]

ID has not proven itself within the scientific community. Because it has failed to gain support within the scientific community, its advocates are attempting to circumvent the scientific community and appeal directly to the public by political means. Some ID proponents recognize the improper nature of this attempt to avoid the rigors of the scientific process, as does Bruce Gordon, an ID proponent and past director of the Baylor Science and Religion Project: "Design theory has had considerable difficulty gaining a hearing in academic contexts, as evidenced most recently by the Polanyi Center affair at Baylor University. One of the principal reasons for this resistance and controversy is not far to seek: design-theoretic research has been hijacked as part of a larger cultural and political movement. In particular, the theory has been prematurely drawn into discussions of public science education where it has no business making an appearance without broad recognition from the scientific community that it is making a worthwhile contribution to our understanding of the natural world."[40]

In view of the critical place that the nature of science holds in building a firm foundation for public science literacy, and in providing a defense against young-Earth creationist and ID arguments, it is incumbent upon the members of the scientific and educational communities to give increased attention to the nature and philosophy of science. Furthermore, regardless of our religious views, we must all finally call a truce in the corrosive and destructive warfare of science and faith. After all, the term *methodological naturalism* was coined for precisely the purpose of describing the universally accessible nature of scientific inquiry.

NOTES

1. This wording from the 1999 Kansas Science Standards Draft 5 (first paragraph under "Nature of Science" section) by the Science Standards Committee was restored to the "Nature of Science" section in the science standards by the newly elected majority of the Kansas Board of Education in February 13, 2007. These standards can be downloaded from the Kansas Department of Education Web site, www.ksde.org.

2. J. Calvert's testimony at the Kansas State Board of Education public forum, July 13, 1999. A complete record of his comments is archived on the Intelligent Design Network's Web site, www.intelligentdesignnetwork.org/7139rem.htm.

3. J. Sjogren, 2000, Idnet Urges Kansas School Boards to Reject National Science Standards Proposed by Kansas Citizens for Science, Intelligent Design Network press release, June 8, www.intelligentdesignnetwork.org/press_releases.htm.

4. Proposed Revisions to Kansas Science Standards Draft 2 with Explanations, 2005, Resources, March 29, www. kcfs.org/kcfsnews. Referred to as the "Minority Report."

5. J. E. Jones III, 2005, Memorandum Opinion, *Kitzmiller v. Dover Area School District*, case 4:04-cv-02688 JEJ, document 342, U.S. District Court for the Middle District of Pennsylvania, December 20, www.pamd.uscourts.gov/kitzmiller/kitzmiller_342.pdf, p. 136.

6. J. Calvert, 2005, Dover Court Establishes State Materialism, Intelligent Design Network press release, December 21, www.intelligentdesignnetwork.org/press_releases.htm.

7. P. E. Johnson, 1999, Response to Denis O. Lamoureux, pp. 49–56 in P. E. Johnson and D. O. Lamoureux, eds., *Darwinism Defeated? The Johnson-Lamoureux Debate on Biological Origins*, Regent College, Vancouver, BC, p. 50.

8. P. E. Johnson, 1991, *Darwin on Trial*, InterVarsity, Downers Grove, IL, pp. 114–15.

9. R. Boston, 1999, Missionary Man, *Church and State Magazine*, April 2. www.au.org/site/News2?abbr=cs_&page=NewsArticle&id=5937&news_iv_ctrl=1037.

10. H. M. Morris and G. E. Parker, 1982, *What Is Creation Science?* Creation-Life, San Diego, CA, p. xii.

11. W. A. Dembski, ed., 1998, Redesigning Science, pp. 93–112 in *Mere Creation: Science, Faith, and Intelligent Design*, InterVarsity, Downers Grove, IL, p. 99.

12. H. M. Morris and G. E. Parker, 1982, *What Is Creation Science?* Creation-Life, San Diego, CA, p. xiii.

13. J. R. Moore, 1979, *The Post-Darwinian Controversies*, Cambridge University Press, Cambridge, UK; D. N. Livingstone, 1987, *Darwin's Forgotten Defenders: The Encounter between Evangelical Theology and Evolutionary Thought*, Eerdmans, Grand Rapids, MI; A. E. McGrath, 1998, *The Foundations of Dialogue in Science and Religion*, Blackwell, Oxford, UK.

14. K. B. Miller, ed., 2003, *Perspectives on an Evolving Creation*, Eerdmans, Grand Rapids, MI; D. R. Falk, 2004, *Coming to Peace with Science*, InterVarsity, Downers Grove, IL; G. L. Murphy, 2003, *The Cosmos in the Light of the Cross*, Trinity Press International, Harrisburg, PA; J. F. Haught, 2000, *God after Darwin: A Theology of Creation*, Westview, Boulder, CO; J. Polkinghorne, 1986, *One World: The Interaction of Science and Theology*, Princeton University Press, Princeton, NJ; D. Edwards, 1999, *The God of Evolution*, Paulist, New York.

15. P. deVries, 1986, Naturalism in the Natural Sciences: A Christian Perspective, *Christian Scholars Review* 15:388–89.

16. Ibid., p. 396.

17. Ronald Numbers indicates that deVries thought that he had coined the term. R. Numbers, 2003, Science without God: Natural Laws and Christian Beliefs, pp. 265–86 in D. C. Lindberg and R. L. Numbers, eds., *When Science and Christianity Meet*, University of Chicago Press, Chicago, IL, p. 320.

18. E. S. Brightman, 1937, An Empirical Approach to God, *Philosophical Review* 46(2): 157–58.

19. P. E. Johnson, 1999, Response to Denis O. Lamoureux, pp. 49–56 in P. E. Johnson and D. O. Lamoureux, eds., *Darwinism Defeated? The Johnson-Lamoureux Debate on Biological Origins*, Regent College, Vancouver, BC, p. 50.

20. W. A. Dembski, ed., 1998, Introduction, pp. 13–32 in *Mere Creation: Science, Faith and Intelligent Design*, InterVarsity, Downers Grove, IL, p. 20.

21. Ibid., p. 27.

22. For a concise discussion of providence and miracle, see J. Polkinghorne, 1989, *Science and Providence: God's Interaction with the World*, Shambhala, Boston, MA.

23. P. Nelson and J. M. Reynolds, 1999, Young Earth Creationism, pp. 39–75 in J. P. Moreland and J. M. Reynolds, eds., *Three Views on Creation and Evolution*, Zondervan, Grand Rapids, MI.

24. W. A. Dembski, 1998, Introduction, pp. 13–32 in *Mere Creation: Science, Faith and Intelligent Design*, InterVarsity, Downers Grove, IL.

25. I have made these arguments elsewhere. K. B. Miller, 1999, Design and Purpose within an Evolving Creation, pp. 109–20 in P. E. Johnson and D. O. Lamoureux, eds., *Darwinism Defeated? The Johnson-Lamoureux Debate on Biological Origins*, Regent College, Vancouver, BC, pp. 109–20; K. B. Miller, 2005, Countering Public Misconceptions about the Nature of Evolutionary Science, *Southeastern Biology* 52:415–27 and simultaneously published in *Georgia Journal of Science* 63:175–89.

26. P. E. Nelson, 1998, Applying Design within Biology, pp. 148–76 in *Mere Creation: Science, Faith and Intelligent Design*, InterVarsity, Downers Grove, IL, p. 149.

27. P. E. Johnson, 1991, *Darwin on Trial*, InterVarsity, Downers Grove, IL, p. 110.

28. P. Nelson and J. M. Reynolds, 1999, Young Earth Creationism, pp. 39–75 in J. P. Moreland and J. M. Reynolds, eds., *Three Views on Creation and Evolution*, Zondervan, Grand Rapids, MI, p. 49.

29. Ibid., p. 50.

30. W. A. Dembski, 1998, *The Design Inference: Eliminating Chance through Small Probabilities*, Cambridge University Press, Cambridge, UK.

31. P. E. Nelson, 1998, Applying Design within Biology, pp. 148–76 in *Mere Creation: Science, Faith and Intelligent Design*, InterVarsity, Downers Grove, IL, p. 150.

32. P. Nelson and J. M. Reynolds, 1999, Young Earth Creationism, pp. 39–75 in J. P. Moreland and J. M. Reynolds, eds., *Three Views on Creation and Evolution*, Zondervan, Grand Rapids, MI, p. 47.

33. M. J. Behe, 1996, *Darwin's Black Box: The Biochemical Challenge to Evolution*, Touchstone, New York, p. 193.

34. Ibid.

35. I have elsewhere stated, "If a person cannot see God in a sunset or a thunderstorm, he or she will not see him in a strand of DNA or a mitotic spindle." K. B. Miller, 1998, God's Action in Nature, *Perspectives on Science and Christian Faith* 50:75.

36. M. J. Behe, 1996, *Darwin's Black Box: The Biochemical Challenge to Evolution*, Touchstone, New York, p. 197.

37. I discuss these various public misunderstandings at length in K. B. Miller, 2005, Countering Public Misconceptions about the Nature of Evolutionary Science, *Southeastern Biology* 52:415–27.

38. P. Nelson, 2004, Interview: The Measure of Design, *Touchstone*, July–August, pp. 64–65.

39. W. Dembski, 2002, keynote address delivered at RAPID (Research and Progress in Intelligent Design) Conference, Biola University, La Mirada, CA, October 25.

40. B. Gordon, 2001, Intelligent Design Movement Struggles with Identity Crisis, *Research News and Opportunities in Science and Theology*, January, p. 9.

EIGHT · *On the Origin of Species* and the Limits of Science

DAVID W. GOLDSMITH

Currently the United States is home to a small but mobilized grassroots effort to have the model of intelligent design (ID) integrated into public school science curricula. What might seem strange is that many of the leading advocates for ID are well aware that by most current definitions, ID does not qualify as science (CSC Top Questions). Many ID advocates admit that their methods and conclusions go beyond what are conventionally accepted as the limits of appropriate scientific explanations; but this, they claim, betrays a flaw in our current conceptions of science. Truth, they say, lies beyond the arbitrary rules of scientific orthodoxy, and only by pushing those boundaries can we find the true nature of the universe. Unfortunately for its proponents, ID makes no compelling claims as to why its methods should be admitted into the fold of legitimate science. In this essay, a comparison of ID with natural selection, a theory that in fact broke new methodological ground in the past, will demonstrate that the exclusion of ID from proper science is due not to some shortsightedness on the part of the scientific community, but to ID's own implicit weaknesses.

ID is not the first theory that has challenged the scientific community to rethink the limits of the scientific method. In some cases, revolutionary discoveries have required scientists to discard everything they previously thought about a subject or its proper method of study. When Charles Darwin first published *On the Origin of Species*, many of his critics seized on his methods just as fiercely as they did on his conclusions. The resulting debate among the scientific community helped to expand the toolbox of scientific investigation. Darwin's book literally changed what it meant for a theory to be considered scientific (Ellegård 1957). In the case of ID, however, the conflict between theory and scientific norms reflects not a need to expand our definition of science but rather a need to clarify the boundaries of science.

A comparison of the intellectual contexts in which natural selection and modern ID were first proposed makes apparent the important differences in their relationships with contemporary scientific thought. Darwin was an integral part of a wider intellectual movement. Victorian philosophers of science, including William Whewell (1840) and John Stuart Mill (1843), were already arguing for a broader range of acceptable scientific methodologies. Darwin was then an exemplar of how these new methodologies could be used to explore nature. ID advocates, on the other hand, offer no independent justification for an expansion of scientific norms beyond promoting their own arguments. The few attempts that have been made to broaden state science standards to include ID are not reflective of a broader intellectual movement. Instead, they are a type of special pleading designed purely for the purpose of legitimizing ID after the fact. Taken to their logical conclusions, the type of pseudoscientific methods that ID adherents advocate would actually weaken the ability of all scientists to posit any acceptable explanations.

ON THE ORIGIN OF SPECIES AND ITS INTELLECTUAL CONTEXT

Perhaps appropriately, Charles Darwin's theory of natural selection did not originate fully formed. The idea that human life may have originated from some earlier ancestral species can be traced all the way back to the Greek philosopher Anaximander in the sixth century BCE. In the late eighteenth and early nineteenth centuries, these notions of transformational biology were even experiencing a mild revival through the works of the French naturalist Jean Baptiste Lamarck and Darwin's grandfather Erasmus Darwin. What made Charles Darwin's work controversial was not simply the ideas that he proposed, but how he defended those ideas.

In publishing *On the Origin of Species,* Charles Darwin not only immersed himself in the scientific debate over life's history, the pattern of diversification of life after its origin from one or a few original forms. He also was engrossed in the philosophical debate over the limits of science. Prior to the Victorian Age, British intellectual tradition dictated that only induction was permissible in scientific explanation. That is, the proper way to understand the world was to examine a full range of particular instances and to extrapolate from these particulars to a general law. Scientific deduction, the formulation of general laws that are then verified through observation, was anathema to most British scientists, who viewed it either as a throwback to the benighted times of Aristotle or as the type of frivolous speculation more typical of French intellectualism (Ellegård 1957).

At the beginning of the nineteenth century, proper British science followed a very rigid methodology. In fact, the scientific method was a point of national pride with two archetypical English heroes. Francis Bacon, the canon claimed, brought the scientific method to a state of maturity by shaking off the Aristotelian commitment to deduction from universal principles. And through his strict adherence to the Baconian method, Sir Isaac Newton, the second great figure of English science, unlocked the secrets of celestial motion.

In the Victorian historiography of science, Francis Bacon was the great emancipator. Aristotle may have begun Western scientific inquiry, but his adherence to deduction from general axioms had severely retarded its progress for centuries. Bacon was among the first to point out the effect that human prejudice and expectation could have on science. In the *Novum Organum* Bacon writes, "The idols and false notions which have already preoccupied the human understanding, and are deeply rooted in it . . . will again meet and trouble us in the instauration of the sciences, unless mankind, when forewarned, guard themselves with all possible care against them" (1620). According to Bacon, human intuition is the weak link in the scientific method. Only through pure and unbiased observation can the true nature of the universe be discovered.

While Bacon describes the philosophical rules for proper science, it is Isaac Newton whose work best exemplifies the use of these rules. In the *Principia*, Newton explicitly states the rules of inference that he considers permissible in scientific inquiry: "RULE IV: In experimental philosophy we are to look upon propositions collected by general induction from phenomena as accurately or very nearly true, notwithstanding any contrary hypotheses that may be imagined, till such time as other phenomena occur, by which they may either be made more accurate, or liable to exceptions. This rule we must follow, that the argument of induction may not be evaded by hypotheses" (1687). According to Newton, only directly observable phenomena are permissible in constructing explanations. For Newton, as for Bacon, the task of the scientist is synthesis—to take disparate observations and induce from these particular cases the general workings of the universe.

Newton demonstrates the extent of his commitment to induction in an amendment to the third edition of the *Principia.* In response to critics who challenge him to explain the cause of gravity, Newton (1687) answers with the now-famous phrase, *"Hypotheses non fingo"* (I make no hypotheses). In claiming unapologetically that he is simply reporting the facts rather than interpreting them, Newton both rebuffs his critics and reaffirms his commitment to contemporary scientific practice. The job of a good Baconian scientist is to induce principles from observations, not to indulge in vain speculation.

Even as late as 1800, the paradigm of science as purely inductive still held firmly, particularly in geology. In 1807 the Geological Society of London held its first meeting and resolved "that there be forthwith instituted a Geological Society for the purpose of making geologists acquainted with each other, of stimulating their zeal, of inducing them to adopt one nomenclature, of facilitating the communications of new facts and of ascertaining what is known in their science and what remains to be discovered" (Geological Society). The society's founders considered accumulation and communication of facts to be proper goals for the society, but not necessarily the explanation of facts. Surveying the state of geology in the early nineteenth century will illustrate both the origin and the significance of this decision.

Prior to the founding of the Geological Society of London, scientists and philosophers who studied the Earth could be characterized as either cosmogonists or geologists. Cosmogonists represented the vast majority of pre-nineteenth-century contemplators of the Earth. They debated the Earth's history based primarily on a desire to demonstrate notions that they held a priori. Actual observation was rarely included in their debates, and when it was, it was simply to illustrate a point, not to form the foundations of a theory (Gillespie 1959). The cosmogonist tradition yielded the famous debate between Neptunists and Plutonists on the origin of granite, and included such pre-Victorian luminaries as Thomas Burnet and Comte Georges Louis Leclerc Buffon.

Those thinkers that we might classify as geologists in the modern sense constituted a minority of Earth scholars in pre-Victorian Europe. They were primarily miners, surveyors, and mineralogists, and their knowledge of the Earth was more practical than that of the cosmogonists. It is this minority group that actually founded the Geological Society of London. Their choice of name for the society is particularly telling. Of the original founders of the society, all but one was a mineralogist (miners and surveyors being initially excluded on social grounds). The choice of the name "Geological Society" rather than "Mineralogical Society" reflects the founders' belief that a central clearinghouse for mineralogical knowledge and description is a crucial stepping-stone toward a more general understanding of the Earth as a whole (Rudwick 1985).

The initial mission of the Geological Society of London "of facilitating the communications of new facts and of ascertaining what is known" also reflects the founders' beliefs of what good science entails. The society would maintain a commitment to induction to improve the state of geology. Previous cosmogonist systems of geology would be considered, by comparison, "a species of mental derangement," according to the premiere edition of *Transactions of the Geological Society* (Fitton 1811, p. 208). This premiere edition reinforces the society's philosophical position by quoting from the *Novum Organum* on its cover (Laudan 1977): "But if any human being earnestly desire to push on to new discoveries instead of just retaining and using the old; to win victories over Nature as a worker rather than over hostile critics as a disputant; to attain, in fact, to clear and demonstrative knowledge instead of attractive and probable theory; we invite him as a true son of Science to join our ranks, if he will, that, without lingering in the forecourts of Nature's temple, trodden already by the crowd, we may open at last for all the approach to her inner shrine" (Bacon 1620). Even at a time when some of the most important theoretical volumes on the workings of the Earth were being published in Germany and Scotland, the Geological Society of London maintained its antitheoretical stance (Rudwick 1985). This policy was not merely some curmudgeonly attempt to curtail wild speculation. It was an adherence to a scientific tradition that stretched back for centuries, upheld by Britain's greatest intellectual heroes.

ON THE ORIGIN OF SPECIES AND ITS LOGICAL STRUCTURE

Darwin's model in *On the Origin of Species* is as contrary to the inductive tradition as is structurally possible (Hodge 1992). Rather than use individual cases to build up to an overarching conclusion, Darwin begins with his model. He then asks a simple question: If this model is correct, what would we expect the world to look like? Darwin acknowledged in his personal correspondence that this mode of reasoning might be dismissed as a type of "what if" storytelling, and anathema to many scientists (F. Darwin 1903). However, he was confident that his was the proper form of argument to answer questions in a historical science.

What is perhaps most remarkable about *On the Origin of Species*, particularly considering its intellectual context, is that it contains no experimentation whatsoever. It is an argument. In the first four chapters, Darwin asserts the adequacy of selection to enact changes in the forms of living things and then supports this assertion through myriad lines of observation. Throughout the remainder of the book Darwin refutes potential counterclaims to his hypothesis and discusses its ramifications, but never proposes an experimental test of his model. Darwin was well aware of the unorthodox structure of his argument, but was also confident that it was the most powerful way to address evolutionary questions. In an 1861 letter to the English botanist J. D. Hooker, Darwin says of one of his reviewers, "he is one of the very few who see that the change of species cannot be directly proved and that the doctrine must sink or swim according as it groups and explains phenomena. It is really curious how few judge it in this way, which is clearly the right way" (F. Darwin 1903, p. 184).

As Darwin pointed out, the realization that a deductive approach was the correct approach for the study of evolution was not widespread among his peers. While many of Darwin's contemporary critics predictably reacted to the perceived theological implications of his work, the deductive roots of his argument often received equal scorn. Scientists, theologians, and even the popular press eagerly pounced on the logical structure of Darwin's argument as a means to undermine it.

Philosophical objections from the scientific community are probably best exemplified by a review written by Adam Sedgwick. Sedgwick, a past president of the Royal Geological Society, mentored Darwin in geology. As one of the men who helped develop the geologic time scale, Sedgwick intimately knew the directional succession of organisms preserved in the fossil record. Nevertheless, on first reading *On the Origin of Species,* Sedgwick sent a letter to Darwin in which he wrote, "parts I read with absolute sorrow; because I think them utterly false and grievously mischievous—You have deserted—after a start in that tram-road of all solid physical truth—the true method of induction" (F. Darwin 1903). Sedgwick was no kinder to Darwin's method in public. In an 1860 letter to the newspaper *The Spectator*, Sedgwick wrote, "I must in the first place note that Darwin's theory is not *inductive,*—not based on a series of acknowledged facts pointing to a *general conclusion,*—not a proposition evolved out of facts, logically, and of course including them. To use an old figure, I look on the theory as a vast pyramid resting on its apex, and that apex a mathematical point." Even for an individual trained in geology with a thorough understanding of the fossil record, the deductive nature of Darwin's model posed an insurmountable roadblock to its acceptance.

It is noteworthy that even those detractors most likely to object to the content and implications of Darwin's argument focused their critiques on his methods. In an 1860 article for *The Quarterly Review*, Bishop Samuel Wilberforce, one of Darwin's best-known contemporary religious critics, specifically concentrated on Darwin's methodological heresies rather than on his ecclesiastical ones. In discussing Darwin's lack of empirical rigor, Wilberforce wrote, "There are no parts of Mr. Darwin's ingenious book in which he gives the reins more completely to his fancy than where he deals with the improvement of instinct by his principle of natural selection. We need but instance his assumption, without a fact on which to build it, that the marvelous skill of the honey-bee in constructing its cells is thus obtained" (p. 253). According to Wilberforce's review, methodological weakness rather than religious objection should have caused Darwin's contemporaries to question his conclusions. Wilberforce claimed that Darwin's conclusions should be embraced despite their apparent contradiction of scripture if they were well supported by experimentation:

> Our readers will not have failed to notice that we have objected to the views with which we have been dealing solely on scientific grounds. We have done so from our fixed conviction that it is thus that the truth or falsehood of such arguments should be tried. We have no sympathy for those who object to any facts or alleged facts in nature, or to any inference logically deduced from them, because they believe them to contradict what it appears is taught by Revelation. We think that all such objections savour of a timidity which is really inconsistent with a firm and well-instructed faith. (1860, p. 256)

Wilberforce's sincerity in stating that his objection to Darwin's ideas was purely methodological must be taken with a grain of salt, however, because later he struggled to squelch them on religious grounds. However, his rhetoric in this review helps to underscore the strength of the methodological objection to Darwin. Wilberforce undoubtedly opposed Darwin's model on religious grounds. However, clearly he felt that attacking natural selection as a hypothesis that lacked experimental evidence would undermine support for it more effectively.

The fact that the scientific community more readily accepted inductively based hypotheses is well illustrated by comparing the reception of the theory of natural selection with that of the theory of ice ages. In the late 1830s, geologists had begun to question old models regarding the origin of erratic boulders found in the U-shaped valley of the Alps. Previous naturalists had explained the presence of these enormous and anomalous rocks through the actions of oceanic currents, icebergs, and even compressed air in underground caverns.

In his 1837 presidential address to the Swiss Society of Natural Sciences at Neuchâtel, the Swiss naturalist Louis Agassiz first proposed glacial ice as a potential source for erratics. His model was met with a mix of skeptical silence and outright hostility. The German naturalist and aristocrat Baron Friedrich W. K. H. Alexander von Humboldt suggested to Agassiz in a letter that by forgetting the entire affair he might "render a greater service to positive geology, than by these general considerations . . . which, as you will know, convince only those who give them birth" (Hallam 1983, p. 71). Agassiz responded to his critics with the publication of a massive treatise on glaciers, highlighting each of the individual observations that had led him to his conclusions (Agassiz 1840). By the mid-1840s, the glacial origin of erratics had become the consensus view among European geologists. When Agassiz had presented his theory as a theory with supporting evidence, it was scorned. When he presented the theory as data with an inductively drawn conclusion, it was embraced.

While modern readers might think that an argument about induction versus deduction and the relative value of experimentation in science might be a debate purely among an academic elite, this was not necessarily true in Darwin's time. Even in the popular press, commentary on Darwin's theory included a critique of method. An 1871 edition of the magazine *Punch* contained the following poem:

DARWIN AND PICKWICK

"Hypotheses non fingo,"
Sir Isaac Newton said.
And that was true, by Jingo!
As proof demonstra*ted*

But Darwin's speculation
Is of another sort;
'Tis one which demonstration
In nowise doth support.

Time, theory's dispeller,
Will out of mind remove it.
We say, as said old Weller,
"Prove it. And he can't prove it."

The fact that even the editorial humor of the day made reference to Darwin's method illustrates how important such considerations were in Victorian England. Scientific method was not merely a point of academic debate; it was a point of national pride. For his heretical departure from Baconian ideals, Darwin found himself pilloried from all quarters.

DEFENDING DARWIN'S DEDUCTIONS

From the preceding critiques of *On the Origin of Species*, one might infer that Darwin was a methodological radical—a maverick whose views had no business being accepted into the fold of "good science." However, by 1859 the traditionalist view of science as a purely inductive enterprise was slowly beginning to change. William Whewell's *The Philosophy of the Inductive Sciences, Founded upon Their History*, published in 1840, and John Stuart Mill's *A System of Logic*, published in 1843, had begun to question the necessity of pure induction and to support a new role for deduction in the scientific method.

Whewell took particular exception to Newton and his infamous claim *"Hypotheses non fingo."* Newton claimed that only those principles induced from direct observation of phenomena had any place in science. Whewell countered this supposition by pointing out that it, in and of itself, is a hypothesis, and a potentially stifling one: "This is, in reality, a superstitious and self-destructive spirit of speculation. Some hypotheses are necessary, in order to connect the facts which are observed, some new principle of unity must be applied to the phenomena, before induction can be attempted" (Whewell 1840). For Whewell, hypotheses and presuppositions were not only necessary to fruitful science; they were inevitable. Only through superimposing some conception upon the facts could induction ever be possible.

If deduction were permitted as a legitimate component of scientific inquiry, then new modes of scientific investigation would become available. Whewell distinguished between two very different ways to do science. The "Colligation of Facts" referred to the Baconian tradition of assembling experimentally derived observations into general laws. The "Consilience of Inductions," on the other hand, was a more theoretically driven mode of inquiry. According to Whewell, a theory could be derived independent of observation and then verified after the fact. Consilience-driven science allowed for the verification of hypotheses if they consistently and simply explained facts observed independently.

For John Stuart Mill, a hypothesis was more than simply a framework in which to view the facts; it was a guide to test inductions. In fact, Mill was more permissive than Whewell when delineating the range of hypotheses that might be allowable in science, but more restrictive in what he considered a proven hypothesis (Ellegård

1957). In discussing the proper method for verifying inductions, Mill wrote, "The hypothesis, by suggesting observations and experiments, puts us on the road to that independent evidence if it be really attainable; and till it be attained, the hypothesis ought only to count for a more or less plausible conjecture. This function, however, of hypotheses, is one which must be reckoned absolutely indispensable in science" (1843, pp. 15–16). Like Whewell, Mill considered hypotheses to be an integral component of the scientific method. Mere reporting of facts does not constitute science. One cannot draw conclusions without a preconceived intellectual framework. In fact, preconceived notions, whether arising from innate qualities of the human mind or unique experiences, make scientific inference possible. Furthermore, such fundamental assumptions also provide the intellectual context for making observations in the first place. They determine which of all possible observations are actually made.

Darwin was well aware that many of his contemporaries would consider a deductively based model to be nonscientific. However, he was also aware of the logical weakness of such objections. In an 1861 letter to the economist Henry Fawcett, Darwin wrote, "About thirty years ago there was much talk that geologists ought only to observe and not to theorise; and I well remember someone saying that at this rate a man might as well go into a gravel-pit and count the pebbles and describe the colours. How odd it is that anyone should not see that all observation must be for or against some view if it is to be of any service!" (F. Darwin 1903, p. 195).

In this sentiment, that theories can suggest observations just as well as observations can suggest theories, Darwin echoed John Stuart Mill. The fact that Mill was both aware and supportive of Darwin's work is evident in an 1861 letter that Fawcett wrote to his friend Darwin: "I was particularly anxious to point out that the method of investigation pursued [in *On the Origin of Species*] was in every aspect philosophically correct. I was spending an evening last week with my friend Mr. John Stuart Mill, and I am sure you will be pleased to hear from such an authority that he considers that your reasoning throughout is in exact accordance with the strict principles of logic. He also says the method of investigation you have followed is the only one proper to such a subject" (F. Darwin 1903). Clearly delighted with Mill's approval, Darwin wrote back, "You could not possibly have told me anything which would have given me more satisfaction than what you say about Mr. Mill's opinion. Until your review appeared I began to think that perhaps I did not understand at all how to reason scientifically" (F. Darwin 1903).

In the century and a half since Darwin's initial publication, his style of deductive argument has been largely accepted as a proper aspect of the scientific method. This is true, at least in part, because Darwin illustrated the power of arguing from a theory. Francis Bacon had redefined the limits of science to exclude deduction in the *Novum Organum*. Whewell and Mill advanced the limits of science by positing that good science could also include speculation, provided that speculation made testable predictions that could then be verified. Within just a few decades of Bacon's writing, his model had its hero in Isaac Newton. In 1859, with the publication of *On the Origin of Species,* Charles Darwin became the Isaac Newton of theoretically driven science.

INTELLIGENT DESIGN AND ITS LOGICAL STRUCTURE

Perhaps appropriately, the general model of ID is an old model that originated largely in its present form and has remained mostly unchanged for the past few thousand years. The basic logical structure of the argument goes back to Aristotle, who argued that while all phenomena have causes, and each of these causes has a cause of its own, there cannot be an infinite regress. Eventually you reach an uncaused cause—the *prime mobile,* or prime mover. The philosopher Thomas Aquinas co-opted this argument into a proof of the existence of God, which was then rephrased by William Paley in the nineteenth century as the well-known "watchmaker" argument from design:

> When we come to inspect the watch, we perceive . . . that its several parts are framed and put together for a purpose, e.g. that they are so formed and adjusted as to produce motion, and that motion so regulated as to point out the hour of the day; that if the different parts had been differently shaped from what they are, or placed after any other manner or in any other order than that in which they are placed, either no motion at all would have been carried on in the machine, or none which would have answered the use that is now served by it. . . . the inference we think is inevitable, that the watch must have had a maker—that there must have existed, at some time and at some place or other, an artificer or artificers who formed it for the purpose which we find it actually to answer, who comprehended its construction and designed its use. (Paley 1802)

Modern ID theory is essentially a quasi-secular reparsing of Paley's argument. Consider the following excerpt from *Darwin's Black Box*, by Michael Behe, a book widely considered to be the founding document of modern ID. In this passage, Behe makes an argument that the complexity of living things (in this case ciliated bacteria) requires the workings of an intelligent designer: "In summary, as biochemists have begun to examine apparently simple structures like cilia and flagella, they have discovered staggering complexity, with dozens or even hundreds of precisely tailored parts. It is very likely that many of the parts we have not considered here are required for any cilium to function in a cell. As the number of required parts increases, the difficulty of gradually putting the system together skyrockets, and the likelihood of indirect scenarios plummets. Darwin looks more and more forlorn" (1996, p. 73). The arguments are isomorphic; that is, they are structurally identical: Complexity requires planning. Planning requires intelligence. Intelligence requires a designer.

One might rightly wonder whether the inference of design should be considered an induction or a deduction. It is worth noting that both rhetoric and law prohibit ID advocates from answering that question ingenuously. ID advocates claim to infer the presence of a designer through rigorous and unbiased examination of nature—almost the definition of induction. However, their inference is based not on what they find in nature, but on their inability to explain what they find in nature. They are basing a positive claim about the universe (the presence of a designer) on negative evidence: a form of argument that has been discounted since the time of Aristotle.

In actuality, ID is a deduction. ID advocates presuppose the existence of a creative agent and then use that presupposition to explain what they see. Unlike Darwin, however, ID advocates cannot explicitly say that their model is deductive. The principle that Darwin was trying to deductively demonstrate was natural selection, a mechanical process rooted in natural law. ID advocates are presupposing an unknowable and potentially capricious intelligence, almost the very definition of a mythic entity. If ID advocates were to concede that their model is deductive, they would essentially be conceding that it is merely a thinly veiled form of theistic creationism.

When making the legal argument for the inclusion or exclusion of ID from public school curricula, the identity of Behe's designer is of paramount importance. Is this all-powerful, unknowable designer God or a god? For the purposes of this essay, however, the identity and attributes of the designer are irrelevant. What is important from a methodological standpoint is the empirical claim that the designer's presence can be inferred at all. This is precisely where Behe and his adherents claim to be breaking new scientific ground. While science traditionally prefers naturalistic and mechanical explanations, ID advocates claim that this should be a starting point rather than an absolute rule.

It is axiomatic in science that a good theory suggests its own tests through its predictions. The best theories are those that explicitly forbid certain phenomena to occur. Science is then the process of holding these predictions and prohibitions up to scrutiny. Data that is inconsistent with the theory requires us to modify that theory. This is the principle of falsifiability (Popper 1963).

As an example, one of the reasons that the theory of plate tectonics is considered a good theory is that it makes predictions about where earthquakes should and should not occur. Consider the following hypothesis that one might make based on the theory of plate tectonics: "Earthquakes occur only as the result of motion along plate boundaries." Every earthquake that occurs in the Aleutian Islands or along the San Andreas Fault adds support to the hypothesis, but no number of them would ever be sufficient to demonstrate conclusively that the hypothesis is correct. Conversely, some earthquakes occur away from plate boundaries, and require additional explanation. These earthquakes, even though they are the minority of earthquakes, refute the hypothesis that earthquakes occur only as the result of motion along plate boundaries. Since these earthquakes are explainable by other means, however, they do not contradict the larger theory of plate tectonics. If geologists were unable to explain these apparently anomalous earthquakes, their occurrence might eventually call the entire theory of plate tectonics into question.

Under modern definitions of science, a theory must be falsifiable to be considered scientific. ID relies on an alternative mode of evidence. The methodology of ID requires verification through an *absence* of evidence and through falsification of alternatives. If one can demonstrate that a particular scientific model has just one alternative, then one can validate that model by falsifying the alternative.

Unfortunately, this mode of argument is based on invalid claims. First among these is that one can dichotomize the world into diametrically opposed alternatives. Few phenomena exist in nature for which only two possible explanations exist. For example, one cannot prove that the Earth is located at the center of the universe simply by disproving that the moon is located there instead. True dichotomies in nature are typically trivial tautologies: It is either raining outside or it is not. To prove the principle of ID using this mode of argument, one would need to change the structure of the argument from "Biological diversity is a result of either Darwinian selection or intelligent design" to "The universe was either designed or it was not." One would then need systematically to disprove that the universe was not designed. Even if such an effort were possible, what then?

The primary boundary of science that ID advocates claim to be pushing is the premium that science places on naturalistic causes. Currently, for a theory to be considered scientific, it must be reliant on natural causes and the conception of obtainable observations that would refute it must be possible. ID advocates would change this cornerstone of science into a suggestion. Under the logic of ID, naturalistic causes are a good starting point; however, if they are demonstrated to be inadequate, then supernatural agency must be considered as the next explanation. This mode of explanation, however, is anathema to modern scientists. Unlike the Victorian reliance on induction, the modern adherence to natural causes is not a mere preference: it is a requirement. To fully understand the ramifications of accepting supernatural explanations in science, it is worth considering in detail exactly the circumstances under which ID theorists would have us consider them.

According to the logic of ID, supernatural causation should be considered for a phenomenon when no known natural cause is adequate to explain that phenomenon. In actuality, there are two different paths that a scientist might choose to take when encountering such an inexplicable phenomenon. The first of these paths is to continue exploring. If there is no *known* natural cause for a phenomenon, then one must dig deeper to reveal a mechanism that is new to science. To follow this path is to assume that scientific inquiry will eventually reveal all of the subtleties of nature.

The second possible path to take when encountering a phenomenon for which no known natural cause exists is to assume that no natural cause exists at all. This is the methodology of ID, which assumes that no *known* cause means no *knowable* cause. To follow this path is to assume that there are real gaps in what we can know about nature and that no amount of inquiry will ever fill those gaps. This argument (sometimes called the God of the Gaps argument) is pessimistic not only in its implications for scientific inquiry but also in its theological implications. Though it assumes that there is a limit to what science can understand about the world, it also relegates God to the role of null hypothesis. That is, if God is there only to explain the gaps in our knowledge, then every new discovery diminishes God's role in the universe.

The God of the Gaps argument actually presents a double-edged sword to its theistic supporters, however. Under this model, closing gaps in our understanding of nature diminishes the role of God. Conversely, every gap in our knowledge that is exposed, and for which a miracle is invoked as an explanation, demonstrates a flaw in God's creation (Van Till 1991). Interestingly, this logic can be traced back to both Paley and Newton. A perfect God should be able to create a universe that is internally coherent. Implying that God must take an active role in such mundane components of the universe as bacterial flagella and amino acid formation does more to denigrate God's craftsmanship than to verify his existence.

Beyond their inherent epistemological pessimism, there is a more important reason to eschew supernatural causes in science. Accepting the role of a potentially capricious unknowable intelligence in one branch of science undermines not just future discoveries in that one field but all scientific knowledge, past, present, and future. Any physical phenomenon is potentially attributable to the action of supernatural agency. Without specific criteria by which to exclude such agency, any scientific investigation becomes pointless. Every phenomenon would have countless, equally valid explanations. Once biologists accept design by some intelligence as a scientific explanation for the origin of the bacterial flagellum, might geologists not also accept the wrath of that same intelligence as an explanation for the eruption of Mount Pinatubo? If inconstant supernatural agency is a permissible explanation in science, then even the most basic experimentation becomes impossible. How can a chemist be sure that the water in her test tube remained water throughout an entire experiment and did not change, if only briefly, to wine?

These are extreme examples, but they are examples that flow naturally from the logic of ID's implied methodology, and therefore demonstrate how this logic undermines the scientific method. All theory building is based on some criteria by which scientists favor one explanation over another. Allowing unknowable agency into the panoply of acceptable causal mechanisms undermines those criteria. A mechanism that has no basis for rejection and that is a viable alternative to every other explanation does not simply push the boundaries of science; it removes them.

THE INTELLECTUAL CONTEXT OF INTELLIGENT DESIGN

In the 1840s, William Whewell and John Stuart Mill helped to pave the way for natural selection by arguing that deductive argument did indeed have a place in the scientific method. In the United States today, there is considerable grassroots effort to have the logical structure of ID accepted as a part of the scientific method. This movement is primarily one aimed at school standards. Advocates for ID want it incorporated into high school science curricula and are willing to rewrite state definitions of science to achieve this end. In 2005, ID advocates achieved a notable victory toward this end in Kansas, when the State Board of Education expanded its definition of science to include ID.

An important distinction to draw between Victorian supporters of Darwin's method and contemporary supporters of the ID method is motivation. Whewell and Mill both wrote in advance of *On the Origin of Species* and were both presumably unaware of its impending publication. While their expansions of logical practice and scientific method validated Darwin's argument, they did so a priori. Darwin provided an example of how an independently justified line of inquiry might bear fruit.

The Kansas State Board of Education, however, redefined science for a very different reason. In 2005 Kansas ratified new state science education standards that included a redefinition of science. Under the new standards, science is no longer defined as "the human activity of seeking natural explanations for what we observe in the world around us." Instead, science is described as "a systematic method of continuing investigation that uses observation, hypothesis testing, measurement, experimentation, logical argument and theory building, to lead to more adequate explanations of natural phenomena" (Proposed Revisions 2004, p. 3). Note that this new definition no longer puts any constraints on what is considered to be an "adequate" scientific explanation.

A 2004 document entitled *Proposed Revisions to Kansas Science Standards Draft 1 with Explanations* clearly illustrates the motivations of ID advocates. In explaining why the emphasis on natural explanations was dropped from the state science standards, the draft report states, "The current definition of science is intended to reflect a concept called methodological naturalism, which irrefutably assumes that cause and effect laws . . . are adequate to account for all phenomena and that teleological or design conceptions of nature are invalid. . . . This can be reasonably expected to lead one to believe in the naturalistic philosophy that life and its diversity is the result of an unguided, purposeless natural process" (Proposed Revisions 2004, p. 4). Clearly this is not a case of scientific practices changing and new inquiry resulting from that change. In this case the definition of science is being expanded a posteriori specifically to allow a direction of inquiry that does not meet conventional methodological standards. There is no intellectual justification for this change beyond the desire to consider ID good science.

The most telling change to the Kansas state science standards, and the one that best illustrates the methodological weakness of ID, comes in a surprising place—a section of the standards entitled "Teaching with Tolerance and Respect." In the new standards, the following two sentences have been removed: "If a student should raise a question in a natural science class that the teacher determines to be outside the domain of a science class, the teacher should treat the question with respect. The teacher should explain why the question is outside the domain of natural sciences and encourage the student to discuss the question further with his or her family and other appropriate sources" (Proposed Revisions 2004, p. 6). As an explanation for this change, the document offers the following: "The parameters defining 'the domain of science' are ambiguous and scientifically controversial, and thus teachers cannot be expected to be able to accurately identify such questions. . . . This provision has previously been identified as a mechanism for suppressing classroom discussion that may conflict with Naturalism or scientific materialism, a philosophy that [some people] contend should not guide science education about origins" (Proposed Revisions 2004). Ironically, this change would not have been necessary without the previous change to the definition of science. The ambiguity cited arises entirely from haphazard tinkering with the meaning of science in the earlier section. Under the old standards, the domain of science was easy to define. Science searched for natural causes for natural phenomena. Rewriting the definition of science to include supernatural agency makes the boundaries of science "ambiguous" and "controversial."

CONCLUSIONS

Had Charles Darwin published *On the Origin of Species* even fifty years earlier, it would have had a significantly more difficult time gaining acceptance. While the strength of his syllogisms and the volume of his evidence may have been the same, his method would have been widely dismissed as pseudoscience by most of the scientific community. It took a separate revolution in the philosophy of science to make the acceptance of Darwin's noninductive model possible.

By any definition, ID can never be considered science. The type of argument that forms the necessary underpinnings of ID runs counter to scientific inquiry in general. When faced with an unknown, the methodology of ID requires scientists and interested others to resign themselves to the possibility of unknowable phenomena rather than to delve deeper to understand the phenomena.

Perhaps the greatest irony of the modern ID movement is the phraseology of its backers. The largest ID think tank in the United States is the Discovery Institute in Seattle. Fellows at the Discovery Institute argue that we should "teach the controversy" in the interest of fostering deeper thinking about scientific issues. In fact, the very logic of ID eschews deep inquiry and discovery in favor of superficial wonder and mystery.

REFERENCES

Agassiz, L. 1840. *Studies on Glaciers*. Repr., 1967, A. V. Carozzi, trans. Hafner, New York.

Bacon, F. 1620. *The New Organon*. www.constitution.org/bacon/nov_org.htm.

Behe, M. 1996. *Darwin's Black Box*. Free Press, New York.

CSC Top Questions. *Discovery.org*. www.discovery.org/csc/topQuestions.php.

Darwin, C. 1859. *On the Origin of Species*. Repr., 1964, Harvard University Press, Cambridge, MA.

Darwin, F. 1903. *More Letters of Charles Darwin*. John Murray, London.

Darwin and Pickwick. 1871. *Punch*, April 8.

Ellegård, A. 1957. The Darwinian Theory and Nineteenth Century Philosophies of Science. *Journal of the History of Ideas* 18(3): 362–93.

Fitton, W. H. 1811. Transactions of the Geological Society. *Edinburgh Review* 19:208.

Geological Society: About Us. *Geological Society*. www.geolsoc.org.uk/template.cfm?name = archives_geolsochistory.

Gillespie, C. C. 1959. *Genesis and Geology*. Harper and Row, New York.

Hallam, A. 1983. *Great Geological Controversies*. Oxford University Press, New York.

Hodge, M. J. S. 1992. Discussion: Darwin's Mode of Argument. *Philosophy of Science* 59:461–64.

Laudan, R. 1977. Ideas and Organizations in British Geology: A Case Study in Institutional History. *Isis* 68(244): 527–38.

Mill, J. S. 1843. *A System of Logic*. Repr., 1874, Harper and Brothers, New York.

Newton, I. 1687. The Principia Mathematica. http://members.tripod.com/~gravitee/.

Paley, W. 1802. *Natural Theology*. Repr., 2006, Oxford University Press, Oxford, UK.

Popper, K. R. 1963. *Conjectures and Refutations*. Routledge, London.

Proposed Revisions to Kansas Science Standards Draft 1 with Explanations. 2004. *Kansas Science 2005*, December 10. www.kansasscience2005.com/ProposedRevisions_KSstandards.pdf.

Rudwick, M. J. S. 1985. *The Great Devonian Controversy*. University of Chicago Press, Chicago, IL.

Sedgwick, A. 1860. Objections to Mr. Darwin's Theory of the Origin of Species. *The Spectator*, April 7.

Van Till, H. J. 1991. When Faith and Reason Cooperate. *Christian Scholar's Review* 21(1): 33–45.

Whewell, W. 1840. *The Philosophy of the Inductive Sciences, Founded upon Their History*. John W. Parker, London.

Wilberforce, S. 1860. Review of *On the Origin of Species*. *Quarterly Review*, July.

PART THREE • ON RELIGION

NINE · Teaching Evolution during the Week and Bible Study on Sunday

PATRICIA H. KELLEY

INTRODUCTION: A DOUBLE LIFE?

I am a geologist who has spent her thirty-year career studying the evolution of fossil molluscs (clams and snails) preserved in sediments up to 80 million years old from the U.S. Gulf and Atlantic coasts. My knowledge of the fossil record in general, and my own paleontological research in particular, has convinced me that life has evolved through time. For instance, I discovered gradual increases in shell thickness over several million years within a number of mollusc fossil species preserved in sediments along the west shore of the Chesapeake Bay (Kelley 1989). This increased thickness was an evolutionary response to shell-drilling predation by carnivorous moonsnails. Thick shells represent good defenses against predators that drill holes in the shells of their victims; individuals with thicker shells survived such predation and passed the trait of thick shells on to future generations through natural selection. Indeed, species that were preyed upon most heavily evolved the most rapidly (Kelley 1991).

Such changes in fossils through time provide strong support for evolution. Life has changed through time, and my own studies and those of other paleontologists indicate that the Darwinian process of "descent with modification" is responsible. In my judgment, evolution is the best scientific explanation for the sequence of fossils found in the world's sedimentary rocks, and I avidly teach this explanation in my paleontology, and other geology, courses during the week.

Sunday morning, however, finds me teaching in a different venue. I am a committed Christian, married to an ordained Presbyterian minister. For most of the past thirty years I have taught the adult Bible study class on Sunday mornings, except for five years in North Dakota when my husband served a three-church rural parish and we attended three worship services every Sunday morning instead. (In fact, in my son's college application essay—for which he chose to write about science and religion—he mentioned that he had attended more church services than there had been Sundays in his life!)

People are often surprised to learn of my double life: paleontologist and pastor's wife. I recall being asked, back when my husband was in seminary and I was a Ph.D. student studying evolution with Stephen Jay Gould, "Don't you and your husband fight all the time?" I would reply, "Not about evolution!" Others, assuming that evolution is incompatible with the biblical view of creation, have asked me how I reconcile what I teach during the week with what I teach on Sunday mornings. In the present essay, as a geologist with a deep religious commitment, I share my perspectives on science, faith, creationism in general, and the intelligent design controversy in particular. I discuss the differences between science and religion, reveal why creationism and intelligent design are religion rather than science, and explain how I reconcile my research and teaching on evolution with my faith.

HOW SCIENCE WORKS

I am convinced that much of the misconception that evolution and faith are incompatible stems from confusion about the differences between science and religion. In part, scientists have been so busy doing science that they have failed to educate the public about what science is and why it must operate the way it does.

Many people assume that science resides in its facts, the pieces of data obtained by observing or measuring the world around us. However, science is not just a collection of facts about the natural world, such as the Grand Canyon is 277 miles long and up to 6,000 feet deep. Instead, science is a tightly integrated set of facts and theories attained in a very specific way, commonly referred to as "the scientific method." One of the best, most succinct descriptions of science was stated in Judge Overton's decision in the 1982 court case *McLean v. Arkansas Board of Education*, which judged unconstitutional the Arkansas law mandating balanced treatment for "evolution science" and "creation science" (Overton 1982). U.S. district judge

John E. Jones III, in the recent *Kitzmiller v. Dover Area School District* intelligent design case, applied this definition as well (Jones 2005).

First, science is guided by natural law, and its explanations are based on natural law; no supernatural explanations may be invoked. In other words, science is not just the study of natural phenomena and features, such as the Grand Canyon. For reasons I will describe later, the explanations of science must be natural as well—an approach often referred to as "methodological naturalism" (e.g., Pennock 1996; Ruse 2001). A scientific explanation for the existence of the Grand Canyon is that it was carved by the Colorado River. The hypothesis that the Grand Canyon was dug by forty thousand angels could not be considered scientific because it invokes the supernatural. Second, the ideas of science must be derived in a specific way, usually known as the scientific method. Data, referred to as "facts," are obtained by observing or measuring natural or experimental phenomena. Then hypotheses are proposed to explain the facts. Ideally, several hypotheses should be proposed for a set of facts. For example, in the case of the Grand Canyon, an alternative scientific hypothesis would be that a glacier carved the canyon.

Predictions are drawn from each hypothesis, either about future events or, in a historical science like geology, about forthcoming observations concerning the record left by past events. Then, based on the predictions drawn from the available hypotheses, further data are collected to test each hypothesis. If the additional data are not consistent with the predictions of a hypothesis, then evaluation is necessary to determine what went wrong. All kinds of things may be wrong: perhaps the prediction drawn from the hypothesis was based on incorrect assumptions, or perhaps the subsequent test was done incorrectly (see also Brown 1986). Where the assumptions, approach, and contradictory data appear to be valid, however, a hypothesis may end up being modified or even rejected (see Murphy 1993 for a good description of this process).

If a hypothesis survives repeated attempts to disprove it and alternative hypotheses are rejected, that hypothesis gains the status of a theory. A theory therefore is a widely accepted hypothesis that has survived repeated testing of its predictions. This usage is much different from that of the vernacular, in which labeling something as "merely a theory" denigrates an explanation to the level of guesswork. Contrary to popular opinion, a scientific theory is a well-tested explanation of facts based on natural law. Nevertheless, although theories explain facts, they are not themselves facts. Theories can never be considered "proven" but by definition must always be open to further testing (see also Murphy 1993). In other words, scientific theories must be falsifiable: there must be some line of evidence conceivable, which if found would disprove the theory.

To return to the example of the Grand Canyon, the alternative scientific hypotheses of river erosion and glacial erosion are testable, even if no one was around to see the Grand Canyon being formed. We can observe the erosional features produced by modern rivers and modern glaciers. We can predict that if the Grand Canyon was produced by river erosion, it would be V-shaped in profile (like modern river valleys) and bear erosional features produced by rivers, such as "potholes" worn into the bedrock by pebbles entrained in eddies. Alternatively, a glacial valley would be U-shaped in profile and its rocks would bear characteristic striations carved into it by rock debris transported by ice. Observations to date support the river erosion hypothesis and falsify the glacial hypothesis—but even well-supported theories must be open to further testing and potential falsification. Thus, because scientific theories can never become "proven facts," the explanations of science are tentative.

In contrast, theories that involve the supernatural are not scientific because there is no way to disprove (falsify) them. The idea that the Grand Canyon was dug by angels is not testable. Angels, or any other supernatural power, do not perform at the whim of humans. Thus their actions cannot be subject to testing; I can think of no way to study scientifically the digging capacity of an angel. The hypothesis is simply not falsifiable: no conceivable evidence would be sufficient to argue against supernatural causation. Lack of preserved angelic shovels or digging marks could easily be argued against in the case of supernatural causation; for example, one could always argue that the angels were careful and took their shovels with them or that supernatural shovels leave no digging marks. Maybe forty thousand angels did dig the Grand Canyon; I can't disprove it, but the possibility is not worth discussing in a geology class.

We see, then, that science is constrained to use only natural explanations, and its conclusions are always open to further testing. Scientific explanations must exclude supernatural causes of any kind, not because scientists are atheists or advocate a materialist worldview (contrary to the writings of Johnson 1990; see the exchange between Johnson and Pennock reprinted in Pennock 2001, pp. 100–107), but because science consists of hypothesis testing, and only natural causes are testable and falsifiable. Science cannot affirm or deny the existence or action of the supernatural; it is simply incapable of addressing questions about the supernatural.

The debate over the extinction of the dinosaurs and their contemporaries provides an excellent example of the way science works. Although various mechanisms of extinction had been discussed for decades, the hypothesis of asteroid impact proposed by Alvarez et al. in 1980 sparked an exciting geological debate that caught the public's fancy. While conducting a geochemical study for other purposes, Alvarez and coworkers discovered a layer of clay enriched in the element iridium at the boundary between Cretaceous (preextinction) and Paleocene (postextinction) sediments. Iridium is rare at the Earth's surface but is enriched in asteroids and other extraterrestrial bodies. The asteroid hypothesis led to further predictions concerning other geological evidence that should be found in conjunction with an impact: "shocked" quartz grains metamorphosed by the impact, glass spherules resulting from the melting of rock followed by sudden cooling, soot from wildfires. The hypothesis was supported by the discovery of all these lines of evidence; however, detractors of the impact hypothesis argued that intense volcanism could also account for this evidence (e.g., Hallam 1987). Geologists also predicted that the impact should have produced a crater, subsequently identified from geophysical data from the Yucatán Peninsula (Hildebrand et al. 1991), an occurrence difficult to attribute to volcanism.

At present, most geologists accept the hypothesis that an impact closed the Cretaceous Period, but even well-supported hypotheses are open to further testing. Thus predictions continue to be drawn from the impact hypothesis and evidence evaluated. Detractors of the hypothesis persist (e.g., Keller 2005), and alternative hypotheses such as volcanism are in turn subjected to their own testing. This is the way that science works. In contrast, no one has proposed that the asteroid was hurled by a vindictive angel who was jealous of God's apparent fondness for dinosaurs. Maybe that's what happened, but it's not a hypothesis that can be tested, and I'm not going to teach it in my Geology 101 class.

Using the criteria presented before, can evolution be considered science? I have sometimes heard it said that evolution can't be studied scientifically because no one was there to see it happen. However, just as the processes that formed the Grand Canyon have left evidence that can be observed today in the canyon's rocks, the process of evolution has left a record in the characteristics of fossil and living organisms. Predictions can be made about future observations of such characteristics, allowing hypotheses about past evolutionary events and processes to be tested. (See also Strahler 1987, p. 15, for a discussion of hypothesis testing in historical sciences.)

A good example of an evolutionary study using prediction and hypothesis testing is the recent find of the fish *Tiktaalik*, reported by Daeschler, Shubin, and Jenkins (2006). Previously known fossils had shown certain similarities between limbed vertebrates (tetrapods) and lobe-finned (sarcopterygian) fish, leading to the hypothesis that tetrapods evolved from a group of sarcopterygians. Based on the age of the oldest known tetrapod fossils and of sarcopterygians with tetrapod-like traits, Daeschler, Shubin, and Jenkins predicted that likely transitional forms should have lived about 375 million years ago. Geological evidence also indicated that the ancestral sarcopterygians inhabited shallow waters, for instance, from deltaic, estuarine, and stream environments—recognizable from sedimentary structures such as ripple marks and cross-stratification. Consequently the team predicted that stream-deposited rocks of this age in Arctic Canada, which then occupied a much lower latitude and offered a warmer environment, would be the best place to prospect for intermediate fossils representing the transition from fish to land tetrapods. As predicted, these rocks yielded a species with scales, fins, and a jaw structure like those in more primitive sarcopterygians, as well as a shortened skull roof, a neck, and a functional wrist joint—in other words, a mixture of fish and tetrapod traits supporting the evolution of tetrapods from sarcopterygian fish (Daeschler, Shubin, and Jenkins 2006).

Clearly evolution, as a testable and potentially falsifiable idea, meets the criteria of science. Indeed, the term *evolution* refers to both a scientific fact and a scientific theory (Thomson 1982). It is a fact that life has changed through time, the simplest meaning of the term *evolution*. I observe such change in the succession of life forms preserved in the fossil record, including in the details of my own paleontological research. But evolution, in the sense of descent with modification from common ancestors, is also a scientific theory, that is, a well-tested explanation of observed facts based on natural law. In my mind the theory of evolution is the only viable scientific explanation for this change in life through time. Evidence for both the fact and theory of evolution is discussed in other essays in this volume and won't be addressed further here. Instead, I will turn next to the issue of creationism and, in particular, of intelligent design.

CREATIONISM AND THE EVOLUTION OF INTELLIGENT DESIGN

Science and religion are different ways of knowing. Through the scientific method, science speaks of the natural world, including how it works and the natural processes by which it came to be as it is today. Religion provides answers to questions of ultimate meaning. Science must exclude the supernatural from its explanations; religion typically embraces the supernatural (e.g., by seeking the divine presence).

Sources of religious knowledge include faith and revelation (i.e., manifestation of the divine to humans), rather than the scientific method. Whereas scientific ideas must be tentative and falsifiable, no such restrictions are placed on religion.

Creationism, including intelligent design, is not science, despite attempts of proponents to argue otherwise. (Note that I do not use the term *creationism* in the broad sense of belief in a creator acting through natural processes, though all explanations involving a creator, no matter how broadly defined, remain supernatural and are not considered scientific.) Biblical creationism, as defined by such authorities as the Institute for Creation Research, includes the belief that all things in the universe were created by God in six literal days as described in the infallible, factual accounts in Genesis (Strahler 1987), which is clearly a religious doctrine. Beginning in the 1960s, several fundamentalist organizations were formed to promulgate the concept that scientific data support the Genesis view of creation (e.g., the Institute for Creation Research, the Creation Science Research Center, and the Creation Research Society; Strahler 1987). These institutions promoted the concept of "scientific creationism," also known as "creation science," as an alternative to evolution—the only alternative in a dualistic scheme that views evolution and the book of Genesis as the two possible, and mutually exclusive, positions on origins.

The tenets of "creation science," as defined in Arkansas Act 590 (Overton 1982, p. 937), include the "scientific evidences and inferences" of relatively recent, sudden creation of the universe and life from nothing; a global flood as responsible for the Earth's geology; and limited change within originally created "kinds" of animals and plants (natural selection and mutation being considered insufficient to produce existing species from an original ancestor). This definition of creation science was pronounced by Judge Overton to be "unquestionably a statement of religion," with its unacknowledged source the book of Genesis (Overton 1982, pp. 937–38). Creation from nothing is the act of a supernatural deity, which in and of itself excludes this concept from the realm of science. However, the methodologies of creation science also fail to meet the criteria of science. As stated in the court decision (Overton 1982), creationists start with a conclusion, the literal wording of the book of Genesis, and attempt to find scientific evidence to support it. Although they attempt to falsify the theory of evolution, their own concepts are not put to similar rigorous testing, nor can they be, since the explanation is supernatural and thus beyond the possibility of testing.

After courtroom defeat of the approach requiring balanced treatment of creation science and evolution science (*McLean v. Arkansas Board of Education* in 1982;

Edwards v. Aguillard in 1987), creationist strategies evolved again. The modern intelligent design (ID) movement developed in the mid-1980s (Numbers 1998) and gathered steam after the 1987 *Edwards v. Aguillard* decision. Major proponents include the lawyer Phillip Johnson, whose writings criticizing the naturalistic basis of evolution advanced the ID movement in the early 1990s; Michael Behe, who has advocated ID from a biochemical standpoint; and philosopher/mathematician William Dembski, who has argued for ID on the basis of information theory. ID attempts to escape issues of constitutionality by not overtly mentioning the activities of a creator. Instead, an unnamed intelligent agent is considered a better explanation than evolution for evidence of design (Behe's "purposeful arrangement of parts"). The basic argument of ID is that Earth's life-forms are too complex, "irreducibly complex" in Behe's parlance, to be accounted for by natural processes. According to Behe, their existence should be attributed instead to an intelligent agent (1996).

Intelligent design was judged in the recent Dover, Pennsylvania, court case to be religious rather than scientific. As stated in Judge Jones's decision, "Although proponents of the IDM [intelligent design movement] occasionally suggest that the designer could be a space alien or a time-traveling cell biologist, no serious alternative to God as the designer has been proposed by members of the IDM. . . . [T]he writings of leading ID proponents reveal that the designer postulated by their argument is the God of Christianity" (Jones 2005, pp. 25–26). Even if one were to argue that an intelligent designer need not be God, expert witnesses in the case made it clear that the designer is supernatural. Witnesses for the defense also testified that the definition of science would have to be expanded to include the possibility of supernatural causation for ID to be considered science. As described earlier, scientists have very good reasons for excluding the supernatural from their explanations; supernatural explanations cannot be tested. In addition, supernatural explanations act as "science stoppers" (Ruse 2001; see also Plantinga 2001). In other words, turning to a supernatural hypothesis when natural processes appear to provide insufficient explanations prevents further seeking of natural explanations. The work of science appears to be done—even though pertinent unexplored natural explanations may be possible. Science stoppers are dangerous; suppose that geologists had stopped looking for a natural cause for earthquakes, assuming them instead to be inflicted by God upon those deserving punishment for their sins? In this case, the science stopper of supernatural cause would have prohibited geologists from linking earthquakes to plate movement, and therefore from being able to predict which areas are most prone to such natural hazards.

ID was also judged not to be science in the *Kitzmiller* case for the following additional reasons (Jones 2005). First, the arguments of ID based on the flawed concept of irreducible complexity rest on the false dualism cited in the *McLean v. Arkansas Board of Education* court case that criticisms against evolution necessarily imply support for ID. Irreducible complexity itself has been argued against but, even if it were sound, it would simply be a negative argument against evolution, not proof of design. The court also considered that arguments against evolution have been countered by the scientific community, but even if they were valid, they would not be support for ID because the two are not mutually exclusive, nor are they the only alternatives. In addition, it was argued that ID has "failed to publish in peer-reviewed journals, engage in research and testing, and gain acceptance in the scientific community" (Jones 2005, p. 89).

THE THEOLOGICAL AND EDUCATIONAL DANGERS OF INTELLIGENT DESIGN

The courts have determined repeatedly that creationism in any of its permutations, including intelligent design, is not science. As a geologist, engaged on a daily basis in doing the work of science, I support fully this determination. The explanations of creationism, including intelligent design, are supernatural and not based on natural law, as is required of science. Creationist ideas are based on faith and not on the scientific method. The conclusions of science are tentative and open to testing; creation science by definition would have to be subjected to the same criteria. Application of these criteria would mean that faith in a creator stands or falls based on the outcome of scientific testing. Personally, I am offended by the idea that faith issues should be open to testing by the scientific method. I don't want my faith to depend on the outcome of a scientific test.

Although science would stand to lose much if explanations such as intelligent design were taught in public school science classes, I believe that religion would stand to lose far more. As stated by Steffen (2006, p. 23), "Proponents of intelligent design are appealing to science to validate faith. . . . That means science becomes the most important—the only!—way to know and understand." As Steffen concluded, intelligent design thus has the effect of subordinating faith to science. I find it ironic that faith should in effect be undermined by a movement attempting "to reverse the stifling dominance of the materialist world view, and to replace it with a science consonant with Christian and theistic convictions" (statement from the Wedge Document, quoted in Forrest 2001, p. 16; see also Forrest and Gross 2004 for discussion of this document laying out the ID agenda).

If an intelligent agent (i.e., God) is invoked to explain phenomena that appear to be too complex for natural explanations, what happens when natural causes are subsequently discovered for such phenomena? This God of the Gaps approach is dangerous: if God is inserted to plug a gap in our natural explanations, and then the gap is narrowed by discovery of satisfactory natural explanations, God becomes squeezed into a smaller and smaller sphere of action. Indeed, this approach of invoking God only in the rare instances of supposed "irreducible complexity" removes God from the role of sustaining everyday life (Raposa 2006).

Like its creation science predecessors, the ID movement feeds the false conception that science and religion are incompatible (see, for example, Plantinga 1991). I consider this view to be dangerous for both science and religion. The erosion of scientific literacy in the United States, which would be furthered by inclusion of ID in the science curriculum, places the United States at a competitive disadvantage compared to other nations and leaves the country ill-prepared to face the public policy decisions the future will require. As stated by Rudin (2006, p. 14), "The teaching of faith-based creationism or ID, its thinly disguised offspring, will significantly weaken both the quality and quantity of science education in America." Bright young minds are being forced to choose between faith and science (Kelley 1999), which can only hurt both science and religion. The choice is unnecessary.

"SAME STORY. DIFFERENT VERSIONS. AND ALL ARE TRUE."

I believe that the Bible is the word of God, and it is authoritative in my life. However, I am not a biblical literalist. I believe not that God dictated every word of the Bible but that the Bible, in its original languages, was written by people who were inspired directly by a supernatural God. They chose the words to communicate what they thought God was communicating to them. Thus the Bible is God's word in human words. In the following paragraphs, I offer examples of the lack of tenability of literalism, particularly as evidenced in the first two chapters of Genesis, along with the understanding of the intentions of the biblical authors based on the results of modern biblical scholarship.

Modern biblical scholarship has convinced me that a truly literal translation of the Bible is impossible. All translations of the Bible must be interpretations, because of problems with the early Hebrew texts. For instance, the early Hebrew manuscripts included no vowels; the oldest consistent text dates from the first century CE, and vowels were added by the Jewish Masoretic scholars only in the ninth and tenth centuries CE, long after the original authorship (Lambdin 1971). The original meaning of the Hebrew text is obscure in many places; my Bible offers alternative translations for many passages. Thus I argue that no translation can be completely literal; all translations must be interpretations to some extent.

Following Old Testament scholar Conrad Hyers (1984), I also argue that, in making these interpretations, we should be as conservative as possible. In other words, we should try to conserve the original concerns and intent of the authors, rather than interpret the texts in the light of our own concerns. So what was the original concern of Genesis? Was it to provide a scientific, factual account of the origins and development of life? I don't believe so, for a number of reasons.

Steinmetz (2005) describes the writings of the third-century biblical scholar Origen, who on theological grounds opposed a literal reading of Genesis as historically accurate. Origen argued that certain "absurdities" in the accounts, such as the existence of light and dark before the creation of the sun, moon, and stars, "were unsubtle hints from God that he wanted the account of creation read in an altogether different way, not as history but as truth 'in the semblance of history' " (p. 27).

This view is supported by the fact that there are two distinct accounts of creation in Genesis (Genesis 1, "In the beginning . . . ," and Genesis 2, the Adam and Eve story). The passages in Genesis 1 and 2, according to biblical scholars (e.g., Hyers 1983, 1984, 1999), date from different times and represent different contexts and concerns. Genesis 1 dates from the sixth century BCE, the time of the Babylonian exile, and is concerned with the origin of the universe, the Earth, and its inhabitants. Genesis 1 has certain similarities to the *Enuma elish*, the Babylonian creation epic, which is divided into seven tablets with certain parallels to the seven days of the Genesis 1 creation story. As described by Hyers (1983), the *Enuma elish* "exalts Marduk, the patron deity of Babylon, as the supreme deity in the Mesopotamian pantheon. Marduk is extolled for rescuing the cosmos from the threat of the goddess of the watery abyss, Tiamat, out of whose womb the first gods had come. He then established, out of the two halves of the slain Tiamat, heaven and earth; sun, moon, and stars; vegetation; animals and fish; human beings. It is this order, and this cosmology, that Genesis 1 most directly approximates."

Biblical scholars thus have argued that the similarities between Genesis 1 and contemporary Mesopotamian creation stories indicate that Genesis 1 was written in repudiation of these rival stories. The purpose of Genesis 1 was to affirm monotheism and undermine the temptations to idolatry that were especially strong during the Babylonian exile. Genesis 1 presents the God of the Hebrews as the omnipotent creator, in contrast to the pantheon of gods of the Babylonians.

Genesis 2 also argues against idolatry, but from a much different context (Hyers 1984). This passage was written at an earlier time, the tenth century BCE, during the reign of King Solomon. The temptations to idolatry during that time stemmed from Solomon's marriage to foreign wives and the royal blessing thus given to worship foreign gods.

The two Genesis accounts differ in many ways (Hyers 1984, 1999; Greenspahn 1983). The name used for God differs; Genesis 1 uses the Hebrew name "Elohim," translated as "God," whereas Genesis 2 employs "Yahweh," or "Lord God." The mode of creation is much different as well. Elohim in Genesis 1 simply commands ("Let there be light") and the universe obeys: "and there was light." In Genesis 2, the creation process is literally hands-on; Yahweh is down in the mud molding creation out of clay. Other details vary as well; for instance, the treatment of water differs in a way that accords with what geologists understand of the environmental context of the stories. In Genesis 1, written during exile on the Mesopotamian floodplain, the problem is an excess of water, from which the Earth must be separated. However, in Genesis 2 the problem is scarcity of water. Yahweh causes a mist to arise and water the Earth, an important concern in the parched setting in which Genesis 2 was written.

The sequence of events during creation differs substantially between the two Genesis accounts. In Genesis 1, the order of creation is light; firmament; Earth and vegetation; sun, moon, and stars; birds and fish; land animals and humans, with male and female simultaneously. In Genesis 2, the sun, moon, and stars are presupposed and their creation is not mentioned. Following the watering of the dry ground, man (Adam) is created, followed by plants, rivers, animals, and birds, and then lastly woman, Eve. (See Hyers 1984, 1999 for the logic, or rather "theo-logic," behind these differences.)

The Hebrew editors who juxtaposed these texts were not troubled by these problems. To me this juxtaposition suggests that the Genesis stories were not intended to be taken as literal scientific accounts. This point is reinforced by the presence of additional, much different creation stories in the biblical books of Proverbs, Job, and Psalms (Greenspahn 1983). Proverbs 3 and 8 speak of a being named "wisdom," personified as a woman, as the first of God's creations, who then assisted God in the work of creation. Job 26 and Psalm 74 allude to a different

Israelite creation story, involving God's calming and dividing the rebellious sea and crushing the snake- or dragon-like beings Leviathan and Rahab. According to Greenspahn (1983), this story resembles a creation myth known from other ancient Mesopotamian cultures, in which a divine battle against the god of the sea (i.e., chaos) spearheads creation.

Thus the Bible contains multiple creation accounts, inconsistent and sometimes contradictory to one another. To the degree that we see these contradictions as problems, we confuse historical accuracy with truth. I am reminded of a scene from the recent Disney film *Pirates of the Caribbean: Dead Man's Chest.* Johnny Depp's character, Captain Jack Sparrow, and his crew seek the advice of the voodoo priestess Tia Dalma concerning the legend of Davy Jones's locker. When one of the crew members offers a different account of Davy Jones's history from that given by Tia Dalma, she retorts, "Same story. Different versions. And all are true." To our modern minds, Tia Dalma's enigmatic statement is more unbelievable than the grotesquely fantastic creatures conjured up by Disney's special effects team to populate Davy Jones's locker.

"Same story. Different versions. And all are true." Our twenty-first-century minds need to disentangle the concepts of fact and truth. Just because a story is not historically accurate, does not mean it is false. As noted by Origen (Steinmetz 2005), some biblical stories are both true and factual and others, like Jesus's parables of the Good Samaritan and Prodigal Son, are true but not factual. Steinmetz (p. 27) describes Origen's thinking as follows: "Was there actually a good Samaritan who helped a Jew wounded by thieves, or a prodigal son who wasted his father's substance in riotous living? Who knows, and even more important, who ultimately cares? The power of the stories is independent of the question of whether they actually happened in time and space."

Nonhistorical accounts can contain deep truths. To believe this does not demean the biblical accounts. Instead, freedom from literalism allows us to perceive deeper meaning in these accounts of creation, meaning beyond the details of what creature was created on what day. As Hyers (1984, pp. 28–29) states, "a literalist interpretation of the Genesis accounts . . . misses the symbolic richness and spiritual power of what is there, and it subjects the biblical materials and the theology of creation to a pointless and futile controversy." Modern biblical scholarship indicates that the biblical creation stories were not intended as historical, factual descriptions of the process of creation. They are statements of faith by the Hebrew people, faith in the one true creator some know as God. I share that faith.

RESOLUTION

So, how do I reconcile my views of evolution with my belief in God as creator? Speaking as a geologist, I conduct my work using the scientific method, and I accept the explanations of modern science, including evolution, for the way the world operates. On the other hand, speaking not as a geologist but as a person of faith, I believe that God used the natural process of evolution as the means of creating. I do not consider creation as an episode confined to six days at the beginning of time. Instead, I view creation as an ongoing process. To me, God's creativity has spanned billions of years, and it's not over yet. As fellow Presbyterian and director of the Virginia Science Museum Walter Witschey has said, "God's creation is continuing. It began long before we were present to contemplate it. More will be revealed in the future" (2006, p. 8).

Like a variety of religious leaders and organizations, from Roman Catholics to Jews to Protestants to Muslims, I am unthreatened by this perspective. Furthermore, I feel bolstered by the many statements from other religious people that support evolution and/or oppose the teaching of "scientific creationism" in public schools. A collection of such statements has been compiled by Matsumura (1995) and includes statements from the American Jewish Congress; Episcopal, Lutheran, Presbyterian, Unitarian, United Church of Christ, and United Methodist denominations; and Pope John Paul II.

To me, this is a powerful view of God, more potent than the view that holds responsible a creator who was active for a few days and then called it quits or one that sees the creator as a tinkerer at the molecular level who intervenes intermittently. I also find it compatible with the evidence presented to me by my scholarly work as a geologist and consistent with the ideas of many other persons of faith. Evolution does not threaten my faith; it gives me a glimpse of the incredible power of God to create.

REFERENCES

Alvarez, L. W., W. Alvarez, F. Asaro, and H. V. Michel. 1980. Extraterrestrial Cause for the Cretaceous-Tertiary Extinction. *Science* 208:1095–1108.

Behe, M. J. 1996. *Darwin's Black Box: The Biochemical Challenge to Evolution.* Free Press, New York.

Brown, H. I. 1986. Creationism and the Nature of Science. *Creation/Evolution* 18: 15–25.

Daeschler, E. B., N. H. Shubin, and F. A. Jenkins. 2006. A Devonian Tetrapod-like Fish and the Evolution of the Tetrapod Body Plan. *Nature* 440:757–63.

Forrest, B. 2001. The Wedge at Work: How Intelligent Design Creationism Is Wedging Its Way into the Cultural and Academic Mainstream. Pp. 5–53 in R. T. Pennock, ed., *Intelligent Design Creationism and Its Critics: Philosophical, Theological, and Scientific Perspectives*. MIT Press, Cambridge, MA.

Forrest, B., and P. R. Gross. 2004. *Creationism's Trojan Horse: The Wedge of Intelligent Design*. Oxford University Press, New York.

Greenspahn, F. E. 1983. Biblical Views of Creation. *Creation/Evolution* 13:30–38.

Hallam, A. 1987. End-Cretaceous Mass Extinction Event: Argument for Terrestrial Causation. *Science* 238:1237–42.

Hildebrand, A. R., G. T. Penfield, D. A. Kring, M. Pilkington, A. C. Zanoguera, S. B. Jacobsen, and W. V. Boynton. 1991. A Possible Cretaceous-Tertiary Boundary Impact Crater on the Yucatan Peninsula, Mexico. *Geology* 19:867–71.

Hyers, C. 1983. Genesis Knows Nothing of Scientific Creationism: Interpreting and Misinterpreting the Biblical Texts. *Creation/Evolution* 12:1–21.

———. 1984. *The Meaning of Creation: Genesis and Modern Science*. John Knox, Atlanta, GA.

———. 1999. Common Mistakes in Comparing Biblical and Scientific Maps of Origins. Pp. 197–206 in P. H. Kelley, J. R. Bryan, and T. A. Hansen, eds., *The Evolution-Creation Controversy II: Perspectives on Science, Religion, and Geological Education*, vol. 5. Paleontological Society Papers. Paleontological Society, Pittsburgh, PA.

Johnson, P. E. 1990. Evolution as Dogma: Establishment of Naturalism. *First Things* 6:15–22. Repr., 2001, pp. 59–76 in R. T. Pennock, ed., *Intelligent Design Creationism and Its Critics: Philosophical, Theological, and Scientific Perspectives*. MIT Press, Cambridge, MA.

Jones, J. E., III. 2005. Memorandum Opinion. *Kitzmiller v. Dover Area School District*. Case 4:04-cv-02688 JEJ. Document 342. U.S. District Court for the Middle District of Pennsylvania, December 20. www.pamd.uscourts.gov/kitzmiller/kitzmiller_342.pdf.

Keller, G. 2005. Biotic Effects of Late Maastrichtian Mantle Plume Volcanism: Implications for Impacts and Mass Extinctions. *Lithos* 79:317–41.

Kelley, P. H. 1989. Evolutionary Trends within Bivalve Prey of Chesapeake Group Naticid Gastropods. *Historical Biology* 2:139–56.

———. 1991. The Effect of Predation Intensity on Rate of Evolution of Five Miocene Bivalves. *Historical Biology* 5:65–78.

———. 1999. An Honors Course on Evolution and Creationism: Teaching Experiences in the Deep South. Pp. 217–25 in P. H. Kelley, J. R. Bryan, and T. A. Hansen, eds., *The Evolution-Creation Controversy II: Perspectives on Science, Religion, and Geological Education,* vol. 5. Paleontological Society Papers. Paleontological Society, Pittsburgh, PA.

Lambdin, T. O. 1971. *Introduction to Biblical Hebrew.* Charles Scribner's Sons, New York.

Matsumura, M. 1995. *Voices for Evolution.* National Center for Science Education, Berkeley, CA.

Murphy, N. 1993. Phillip Johnson on Trial: A Critique of His Critique of Darwin. *Perspectives on Science and Christian Faith* 45(1): 26–36. Repr., 2001, pp. 451–70 in R. T. Pennock, ed., *Intelligent Design Creationism and Its Critics: Philosophical, Theological, and Scientific Perspectives.* MIT Press, Cambridge, MA.

Numbers, R. L. 1998. *Darwinism Comes to America.* Harvard University Press, Cambridge, MA.

Overton, W. R. 1982. Creationism in Schools: The Decision in McLean vs. the Arkansas Board of Education. *Science* 215:934–43.

Pennock, R. T. 1996. Naturalism, Evidence, and Creationism: The Case of Phillip Johnson. *Biology and Philosophy* 11(4): 534–49. Repr., 2001, pp. 77–98 in R. T. Pennock, ed., *Intelligent Design Creationism and Its Critics: Philosophical, Theological, and Scientific Perspectives.* MIT Press, Cambridge, MA.

———. 2001. *Intelligent Design Creationism and Its Critics: Philosophical, Theological, and Scientific Perspectives.* MIT Press, Cambridge, MA.

Plantinga, A. 1991. When Faith and Reason Clash: Evolution and the Bible. *Christian Scholar's Review* 21(1): 8–32. Repr., 2001, pp. 113–46 in R. T. Pennock, ed., *Intelligent Design Creationism and Its Critics: Philosophical, Theological, and Scientific Perspectives.* MIT Press, Cambridge, MA.

———. 2001. Methodical naturalism? Pp. 339–61 in R. T. Pennock, ed., *Intelligent Design Creationism and Its Critics: Philosophical, Theological, and Scientific Perspectives.* MIT Press, Cambridge, MA.

Raposa, M. 2006. Finding the Divine in the Everyday. *Lehigh Alumni Bulletin,* Winter 2006, pp. 21–22.

Rudin, A. J. 2006. Guest Viewpoint: Don't Teach Religion in Science Classes. *Presbyterian Outlook* 188(8): 14.

Ruse, M. 2001. Methodological Naturalism under Attack. Pp. 363–85 in R. T. Pennock, ed., *Intelligent Design Creationism and Its Critics: Philosophical, Theological, and Scientific Perspectives.* MIT Press, Cambridge, MA.

Steffen, L. 2006. The Courage of Faith. *Lehigh Alumni Bulletin,* Winter 2006, pp. 22–23.

Steinmetz, D. C. 2005. Creator God: The Debate on Intelligent Design. *Christian Century* 122(26): 27–31.

Strahler, A. N. 1987. *Science and Earth History: The Evolution/Creation Controversy.* Prometheus Books, Buffalo, NY.

Thomson, K. S. 1982. Marginalia: The Meanings of Evolution. *American Scientist* 70:529–31.

Witschey, W. R. T. 2006. Intelligent Design: A Cultural Code Phrase. *Presbyterian Outlook* 188(8): 8, 11.

TEN · The "God Spectrum" and the Uneven Search for a Consistent View of the Natural World

WARREN D. ALLMON

The test of a first-rate intelligence is the ability to hold two opposed ideas in mind at the same time and still retain the ability to function.

F. SCOTT FITZGERALD, "The Crack-Up," *Esquire* magazine, February 1936

God, the swamplands we're willing to wade through to get around the truth!

NEIL LABUTTE, *Wrecks*, 2006

Bigotry is an incapacity to conceive seriously the alternative to a proposition.

G. K. CHESTERTON, *London Daily News*, 1910

PREFACE

There was a time not so long ago when scientists did not take creationism seriously. As copiously demonstrated, however, in the flood of recent scientific critiques of intelligent design (ID), including in the other chapters of this volume, this is no longer the case. These responses have exhaustively demonstrated that creationism, including ID, isn't science; it's religion. And for most scientists, the task stops there. Yet for the great majority of the nonscientific public, this is just the beginning. For most people whose lives are primarily occupied by neither science nor religion, the bright line that the majority of scientists see as separating the two is fuzzy to nonexistent. Explicating the difference is therefore sometimes the major intellectual hurdle that any scientist, and especially any science teacher or professor, must clear—even if only perfunctorily—before his or her audience can begin to grasp the details of the geology or evolutionary biology that most of us would much rather talk about.

My experience is that the great majority of my scientific colleagues, perhaps especially geologists, are deeply uncomfortable discussing the details of the differences between science and religion. They feel that they are not qualified, or that it's personal, or irrelevant, or inappropriate, or that they just don't have time with everything else that they have to cover in a semester or during a crowded office hour. Most of us do not explicitly engage this issue, and this, in my view, is a serious mistake. By avoiding this discussion, we avoid what (if polls are to be believed) a huge proportion of Americans hold much more dear than our precious science: religious faith. Trying to ignore religion as we attempt to communicate—and generate support for—our various scientific fields is something like trying to ignore traffic as we attempt to drive across a big city: it just can't be done. It's there, and we must deal with it.

This chapter is an attempt both to deal with religious belief in the context of the manifest practical success of modern geological and evolutionary science and to deal with science in the context of the manifest popularity of religious faith. It is my attempt to answer the question that most professors who have taught historical geology or evolution have confronted or will eventually confront: "Can one believe in both God and evolution?"

INTRODUCTION

Science and religion are the realms of human life "that most fundamentally tell us who we are and define our relationship with the rest of the world" (Croce 1995, p. 16). Yet the relationship between these two huge spheres of human experience has never been the subject of anything approaching a consensus, among either scholars or laypeople. This has been true for more than four hundred years, since the birth of modern science, and it continues to be true today, even with an explosion of scholarly interest in the relationship, marked by the appearance of numerous publications, conferences, and organizations, and even the emergence of a formal field of religion-science studies.

Barbour (1990) proposed a useful, although perhaps overly simplistic (see chapter 9; Bryan 1999), four-fold taxonomy of views of the relationship between science and religion: conflict, independence, dialogue, and integration. The first category has a long history; science and religion have for many years been said to be in "struggle," "conflict," and even "war" (e.g., Draper 1874; White 1896; Ruse 2001b, 2005). Although rejected today by most historians of science (see, e.g., Wilson 2000; Lindberg and Numbers 2003), this point of view is carried on, ironically, by two groups who agree on little else: scientific/materialistic atheists and creationists both hold that the belief in God and the pursuit of modern science (at least major parts of it, particularly but by no means limited to evolution and Darwinism) are utterly irreconcilable.

There is also a long history of what Barbour (2000) calls the "dialogue and integration" (what I will herein call "accommodationist"[1]) view of the relationship between science and religion (see, e.g., Wilson 2000; Bowler 2001; Ferngren 2002; Lindberg and Numbers 2003; Olson 2004; Witham 2005; Thomson 2005), a tradition carried on today by a large and growing number and variety of viewpoints, from the scientific, philosophical, and religious (e.g., Haught 1995, 2000, 2003, 2006; Brooke and Cantor 1998; Goodenough 1998; McGrath 1998; Polkinghorne 1998, 2005; Raymo 1998, 2005; K. R. Miller 1999; Rolston 1999, 2006; Griffin 2000; Knight 2001; K. B. Miller 2003; Kitcher 2007) to the political (e.g., Wallis 2005; Balmer 2006; Lerner 2006; Meyers 2006; Obama 2006).

Moving past this tradition of accommodation in the religious direction, we encounter the long-standing views that God must come first and that science must adjust itself to religion (not the other way around) and to the continual supernatural involvement of an all-powerful deity in the physical universe. In the United States, these views are most manifest today as modern creationism, including its most recent incarnation, intelligent design (e.g., Pennock 1999; Forrest and Gross 2004; Perakh 2004; Scott 2004; Shanks 2004; Young and Edis 2004; Jones 2005; Ayala 2006; Brockman 2006; Numbers 2006; Shermer 2006; Petto and Godfrey 2007; this volume). On the atheist side are authors who argue that science essentially demands that there is, in fact, no God at all (e.g., Provine 1987, 2006; Weinberg 1993, 1999, 2007; Dennett 2006; Dawkins 1986, 2003, 2006; Tyson 2001; Graffin 2003; Harris 2004, 2006; Atkins 2006; Wolf 2006).

At least in the United States, advocates of both extreme opinions claim that they are swimming against a strong current that threatens to overwhelm their preferred worldview. Secularists point to the rise of evangelical and fundamentalist Christianity and the overwhelming self-proclaimed religiousness of the American population. Religionists point to the dominance, particularly in the popular media and academia, of secular and atheistic views. The truth is that the United States today is awash in *both* secularism and religion—a surprising situation that Marty (2006) has labeled "religiosecular"—and American society is in important ways deeply conflicted about it (e.g., Turner 1985; Silk 1988; Lacey 1989; Lachman 1993; Fowler et al. 2004; Jacoby 2004).

Possible positions on the issue of the relationship between science and religion are, however, "many more than usually considered" (Pigliucci 2000, p. 38). Although adequate reliable data are unfortunately lacking, it is clear that the majority of individual Americans—both scientists and nonscientists—hold neither extreme point of view, but rather views that are somewhere along a spectrum in-between: *the majority of scientists, students, and the general public in America hold opinions that combine both science and religion.*

SOME DEFINITIONS

Before delving further into this wide range of views, I want to be clear about how I will be using certain terms.

Religion: The term *religion* is "notoriously difficult to define," as evidenced by the huge diversity of definitions that have been offered (e.g., Smith 1962, p. 17; Abernethy and Langford 1968, p. 1; Brooke 1991; Peterson et al. 2003, p. 6). Part of the problem may be the attempt to group together under one word or concept such an enormous diversity of human ideas, beliefs, and behaviors. The idea of religion, applicable to all of humanity, as distinct from individual religious faiths or traditions, may be in whole or large part a creation of modern (Western) thought; the numerous traditions generally referred to as religions of the world may, in fact, not have enough to unite them easily or usefully under a single term (Smith 1962), at least for the purposes of the discussion in this chapter. When I use the term *religion* here, therefore, it will be in reference only to the three (mono)theistic, Abrahamic faiths (Judaism, Christianity, and Islam),[2] which have in common the belief in the existence of a single divine, supernatural, nonmaterial, ultimate reality that interacts with the material world, especially with humanity; a "transcendent spiritual being who is omnipotent, omniscient, and perfectly good" (Peterson et al. 2003, p. 9).

Materialism/naturalism: Although these two words are not strictly synonymous (see, e.g., Davis and Collins 2000; Jammer 2003; Drees 2003; Flanagan 2006), I will use them interchangeably here. Two different senses of materialism/naturalism are frequently recognized: "methodological" materialism/naturalism holds that science can and should deal only with material causes for material phenomena; "philosophical" (also known as "metaphysical," "theoretical," "hard," "ontological," or "imperialistic ontological") materialism/naturalism holds that the material world is all that exists and denies the existence of anything else. The distinction between these two senses is important, not least because it is widely maintained that it is possible and reasonable to accept the first without accepting the second (see chapter 7; Pennock 1999, p. 189ff.; Davis and Collins 2000; Forrest 2000; Numbers 2003; Scott 2004, pp. 65–68; Flanagan 2006; Rolston 2006).

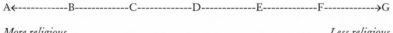

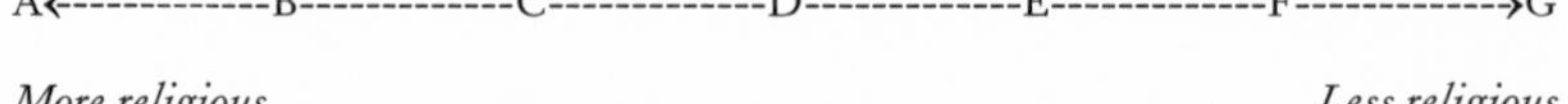

More religious *Less religious*

A. A supernatural (nonmaterial) God designed and created the entire material universe and has maintained and guided it through all time by continuous or frequent specific intervention at any time he/she/it so desires and in any manner, including suspension of known physical laws.

B. A supernatural (nonmaterial) God designed and created the entire material universe and has maintained and guided it through all time by continuous or frequent specific intervention, but this intervention occurs in accord with known physical laws (e.g., quantum indeterminacy).

C. A supernatural (nonmaterial) God designed and created the entire material universe, including physical laws, but since then has let those laws govern the universe and has not physically intervened.

D. A supernatural (nonmaterial) God may or may not have designed and created the entire material universe but is somehow immanent in that material universe, including in living things, and necessary for its continued existence.

E. A supernatural (nonmaterial) God exists, but not in any measurable material sense or fashion that materialistic science could ever detect or describe, and this God communicates in some identifiable way (consciously or unconsciously) with human beings.

F. Some kind of supernatural (nonmaterial) entity or phenomenon may exist but, if so, humans may or may not ever know this with certainty nor ever know with any confidence anything specific about it or its activity or effects.

G. There is no supernatural (nonmaterial) God or anything else that is other than the material universe of matter and energy.

FIGURE 10.1
"The God Spectrum" as discussed in this chapter.

THE SPECTRUM OF BELIEF AND ITS IMPLICATIONS

The reality and population of this spectrum of belief, which I will herein call the God Spectrum (figure 10.1), has major implications for those of us who seek to support science education in general, and to promote and teach historical geology and evolution in particular. This is the point of departure for this chapter. I am concerned here with two separate but related topics. First is the question of the nature of nature—what exists and what doesn't, what its properties are, and what happened or didn't in the past—what Pennock (1999, p. 40) calls "the truth of nature and the nature of truth." Second is the question of how scientists can and should communicate with the public about these topics. I write from my point of view as both a practicing research scientist and an educator and museum director. I want both to understand what the universe is really like, *and* learn how I can most effectively share scientific understanding with others, from preschoolers to college students to senior citizens. I am acutely aware that these perspectives are not always easy to reconcile or amenable to the same kinds of solutions. Although one could obviously present this treatment in a variety of ways, my approach throughout this chapter is to address the issue of education and communication, and let insights about the nature of reality emerge from there.

Although it is widely believed that American academia is currently dominated by secular, "liberal" views (e.g., Marsden and Longfield 1992; Marsden 1996; K. R. Miller 1999; Roberts and Turner 2000), the American public—including the student bodies of most colleges and universities—is clearly overwhelmingly religious. Thus, when academic scientists teach, speak, and write publicly about science in general, and evolution and Earth history in particular, we frequently encounter questions about whether these areas are compatible with religion (e.g., Scott 1996, 1999; Kelley 1999, 2000). In my experience, most teachers and communicators of science to the public respond to such questions in one of three ways: (1) a few of us say that the methods of science and religion are utterly and irreconcilably different and that intellectual honesty requires that one must choose between them; (2) many say that science and religion are not mutually exclusive and can in principle be accommodated, but we do not explain this accommodation in much detail; or (3) most avoid the issue entirely.

If we choose the first option, most of our students and audience will almost certainly select religion rather than science, and this is increasingly dangerous from the point of view of trying to teach and promote science. If we choose the second option—because we think (or hope) either that it's true or that our listeners will thereby be more likely to accept and learn the science—the potential problems are several. Although religion and science may not necessarily be mutually exclusive, it is, however, the case that many practicing scientists who say they are personally religious are not usually very clear or consistent about the implications of holding both outlooks. Furthermore, *not all religious beliefs are consistent or compatible with science,* and so a broad and unqualified statement of accommodation is, in fact, inaccurate. If we choose the third option, we are simply being intellectually lazy and not doing our jobs.

I would like to propose that there is a fourth response. We can say that science and religion are both important approaches that humans use to understand the world, and that there *are* views, held by both respectable theologians and respectable religious scientists, that *God exists and acts in real but nonmaterial ways that are not accessible to or relevant to scientific study of the material world.* In this direction lies an avenue of potentially genuine accommodation between religion and science. This fourth response suggests that there may exist a separate, nonmaterial reality, utterly and permanently inaccessible to the methods of science, and that humans access and interact with this separate reality in ways that scientists, *when working as scientists,* do not. Human understanding of this separate reality can be pursued only by the tools of faith and theology, and perhaps—like the conclusions of science—will always be incomplete and provisional. The existence of this separate reality, however, does not impinge on material reality, for the understanding of which naturalistic science is clearly the best means at our disposal.

If we choose this fourth option, and explain that there *are* legitimate religious views that allow for a supernatural deity *and* are not inconsistent with materialistic science and we explain how, then this may increase the likelihood that our audiences will be able to hear what we say about science in general, and evolution and historical geology in particular. This fourth choice, however, has some major caveats. First, we may not believe it ourselves. Yet, importantly, this does not mean that we can prove scientifically that it is *not* correct. Second, these scientifically acceptable religious views may not grant legitimacy to all or even most of the beliefs of most religious persons. Most prominently, it is difficult, if not impossible, to reconcile any notion of "miracles" with a defensibly consistent scientific worldview. Lastly, this approach may come perilously close to telling students what they can believe in order to do science.

In this chapter I argue that very few of the positions on the God Spectrum—neither those on the ends, nor those in the wide middle—have been explained by their advocates in what I would call sufficiently clear or consistent detail. That is, these individuals have largely failed to clearly identify and resolve potential conflicts or inconsistencies between materialistic and religious viewpoints. Some, perhaps many, people do not see this lack of logical consistency as a problem. I do. I also try to assess what the population density may be along this spectrum of belief—among both the general public and scientists—and I analyze opinions held by scientists, theologians, and philosophers whose views fall at various points along the spectrum. Finally, I offer some modest suggestions for how evolutionists and historical geologists, and particularly educators in these subjects, might proceed.

SEPARATE BUT EQUAL: SCIENCE AND RELIGION AS A NONCONTINUUM

Before examining the evident continuum of views linking naturalistic science and religious belief in the supernatural, I first need to consider whether such a continuum can exist at all, that is, whether science and religion are even addressing the same subjects or whether they are completely distinct.

The idea of science and religion as separate but equal realms of human experience—labeled by Gould (1997, 1999) as "non-overlapping magisteria," or NOMA—holds that neither can nor should make claims on the other's legitimate domain of influence. "No scientific theory, including evolution," wrote Gould, "can pose any threat to religion, for these two great tools of human understanding operate in complementary (not contrary) fashion in their totally separate realms: science as an inquiry about the factual state of the natural world, religion as a search for spiritual meaning and ethical values" (2001, p. 214). Gould clearly saw this as a very important social issue: "People of goodwill wish to see science and religion at peace, working together to enrich our practical and ethical lives" (1999, p. 4). "[T]he myth of a war between science and religion remains all too current," he wrote, "and continues to impede a proper bonding and conciliation between these two utterly different and powerfully important institutions of human life. How can a war exist between two vital subjects with such different appropriate turfs—science as an enterprise dedicated to discovering and explaining the factual basis of the empirical world, and religion as an examination of ethics and values?" (1995, pp. 48–49). We need science to do what it does, he argued, but "we will also need—and just as much—the moral guidance and ennobling capacities of religion, the humanities, and the arts, for otherwise the dark side of our capacities will win, and humanity may perish in war and recrimination on a blighted planet" (2001, p. 269).

> Science can supply information as input to a moral decision, but the ethical realm of "oughts" cannot be logically specified by the factual "is" of the natural world—the only aspect of reality that science can adjudicate. . . . I

> win my right to engage moral issues by my membership in *Homo sapiens*—a right vested in absolutely every human being who has ever graced this earth, and a responsibility for all who are able. If we ever grasped this deepest sense of a truly universal community—the equal worth of all as members of a single entity, the species *Homo sapiens,* whatever our individual misfortunes or disabilities—then Isaiah's vision could be realized, and our human wolves would dwell in peace with lambs, for "they shall not hurt nor destroy in all my holy mountain." We are freighted by heritage, both biological and cultural, granting us capacity both for infinite sweetness and unspeakable evil. What is morality but the struggle to harness the first and suppress the second? (Gould 1995, p. 318)

NOMA, however, did not fare well among theologians, philosophers, or evolutionists (see, e.g., Ruse 1997; Goodenough 1999; Orr 1999; Pigliucci 1999; Coyne 2000; Haught 2000, 2003; Watson 2000). One important reason for this less-than-enthusiastic reception lay in Gould's definition of religion. To have religion not conflict with science, said critics, Gould had to define religion in a way that excluded much of what religious people value, namely the actual existence and action of supernatural forces. To make NOMA work, said theologian John Haught, for example, Gould had to "first reduce 'religion' to ethics" (2000, p. 25). Gould could reconcile science and religion, Haught later elaborated, only

> by understanding religion in a way that most religious people themselves cannot countenance. Contrary to the nearly universal religious sense that religion puts us in touch with the true depths of the real, Steve denied by implication that religion can ever give us anything like reliable knowledge of *what is*. That is the job of science alone. As far as Steve was concerned, our religious ideas have nothing to do with objective reality. Scientific skeptics may appreciate religious literature, including the Bible, for its literary and poetic excellence. But they must remember that only science is equipped to give us factual knowledge. Doubters may enjoy passages of Scripture that move them aesthetically, or they may salvage from religious literature the moral insights of visionaries and prophets. . . . Still, Steve could not espouse the idea that religion in any sense gives us truth. No less than Dennett and Dawkins, when all is said and done, he too held that only science can be trusted to put us in touch with what is. At best, religion paints a coat of "value" over the otherwise valueless "facts" disclosed by science. Religion can enshroud reality with "meaning," but for Steve this meaning is not intrinsic to the universe "out there." It is our own creation. (2003, pp. 6–7)

Ultimately, and ironically, NOMA failed to convince because it was an attempt to do what Gould himself consistently criticized in others: make reality match our hopes. His family background and intellectual leanings made him a nonbeliever in religion, but his cultural heritage imbued him with a deep appreciation of the value of many aspects of religion. His abiding humanism compelled him to seek and find a personal reconciliation of science and religion, but the religion that he thought could coexist in such equality with science is a religion that few believers accept (see Allmon 2009, for further discussion).

THE GOD SPECTRUM

By God Spectrum (figure 10.1), I mean the variety of belief in the amount of direct involvement ("causal commerce"; Flanagan 2006, p. 433) of a supernatural deity in the day-to-day operations of the physical universe, from the extreme supernatural end—in which a supernatural deity can and does do anything at any time, regardless of the laws of nature—to the extreme materialistic end—in which there is no God or supernatural deity.

At the supernatural end is the belief that God has always been closely and intimately involved in all aspects of the physical world—in the case of evolution, for example, he has been creating as he goes. Mechanistically, this can be as supernatural and filled with miracles and divine purpose as one would like. As we move away from this end of the spectrum, however, we encounter the views of many self-described "liberal to moderate" Christians (including at least some practicing scientists), that God has worked mainly through natural laws, suspending them only every once in a while for his own purposes. (I like to think of this point of view as similar to the role of most parents when their child goes to college. The child is on his or her own in some respects, but not completely. The parents don't love the child in just a passive way; they help out in significant ways.) As we move still farther along the spectrum, we eventually cross a line. On one side is an active God; on the other an inactive God. Some people believe that there is a God who created all we see around us, but who has not been involved with the physical world in any significant way since that creation ended. (This might be thought of as similar to the role of most parents when the child is an adult and living and working elsewhere; they still love their child and that makes the child feel good, but they no longer send money.) Somewhere in this area of the spectrum is the view that God exists but science simply can have no opinion on the matter. This may be the closest to what Darwin himself believed at the end of his life.

Other authors have proposed frameworks similar to figure 10.1 to communicate the range of views about the relationship between science and religion. Scott (1997; 2004, p. 57) presents perhaps the fullest version of such a "continuum of religious views." Dickerson (1990) similarly rejects the "radical dichotomy" between science and religion, which is limited to consideration of only two alternatives, in favor of a spectrum of views. My emphasis here, however, is more on the nature of God's *activity* than on the associated belief system(s), and on analysis may thus be most similar to the "Fuzziness of the God concept" of Pigliucci (2002, p. 180). In a similar vein, the anthropic principle has been described by Barrow and Tipler (1986) as less a single perspective and more an array of views about the role of an intelligent designer in making the universe habitable. Dawkins (2006, pp. 50–51) also describes a "spectrum of probabilities" for the existence of God. His spectrum, however, is difficult to apply to the issues I address here because (although he does refer to "versions" of his "God hypothesis"; e.g., 2006, p. 58) Dawkins appears to insist that God of a fixed sort either exists or does not, and leaves little room for discussion of various different ways in which God might act in the world, ways that fall somewhere along the spectrum shown in figure 10.1.

COMPLETE NATURALISM: "THE GOD DELUSION"

The extreme materialistic end of the God Spectrum ("G" in figure 10.1) has a long history, much of which is frequently described as the steady increase in secularity of modern society. Although it is certainly true that numerous predictions about the imminent disappearance of religion have proven to be incorrect (e.g., Berger et al. 1999; Stark 1999; McGrath 2004), it is also true that, at least in most Western countries across the industrialized world, organized religion today has less direct impact on the daily lives of people, governments, and institutions than it did a century and a half ago (e.g., Chadwick 1993; Bruce 2002). This change, whether it is called "secularism" or "secularization" (e.g., Hollinger 1989; Stark 1999), while neither inevitable nor irreversible, is an undeniable long-term trend in Western society, which has accelerated—albeit unevenly—over the past hundred years (Hollinger 1989; Croce 1995; Jacoby 2004; Weinberg 2007). Although clearly not all of this change was caused by science (e.g., Numbers 2003), the trend is seen by many historians as a result of the success and accompanying rise in social status of science. Furthermore, ever since the very beginnings of modern science, some (although certainly not all) of its advocates have in fact also been forceful advocates for antireligious viewpoints (Hollinger 1989; Numbers 2003; Thomson 2005).

"God," says Richard Dawkins, who is perhaps the most vocal and well-known contemporary advocate of this view, is "a pernicious delusion" (2006, p. 31). "Science is the only path to understanding," writes chemist Peter Atkins. "It would be contaminated rather than enriched by any alliance with religion" (Atkins 2006, p. 124). "You clearly can be a scientist and have religious beliefs," says Atkins. "But I don't think you can be a real scientist in the deepest sense of the word because they are such alien categories of knowledge" (quoted in Larson and Witham 1998, p. 313).

Historian of evolutionary biology William Provine similarly argues that it is simply impossible to accept both a traditional, personal God and Darwinism:

> It is still possible to believe in both modern evolutionary biology and a purposive force, even the Judeo-Christian God. One can suppose that God started the whole universe or works through the laws of nature (or both). There is no contradiction between this or similar views of God and natural selection. But this view of God is also worthless. Called Deism in the seventeenth and eighteenth centuries and considered equivalent to atheism then, it is no different now. A God or purposive force that merely starts the universe or works through the laws of nature has nothing to do with human morals, answers no prayers, gives no life everlasting, in fact does nothing whatsoever that is detectable. In other words, religion is compatible with modern evolutionary biology (and indeed all of modern science) if the religion is effectively indistinguishable from atheism. (Provine 1987, pp. 51–52)

Philosopher Daniel Dennett agrees, and argues that accommodation of Darwinism with religion is impossible because they are inherently irreconcilable. If we had the courage to look squarely at what Darwinism really tells us about the way life works, Dennett insists, we would see that any notion of divine influence in nature is "cognitionally empty." Those biologists "who see no conflict between evolution and their religious beliefs" are refusing to face incontestable scientific facts (Dennett 1995, p. 310).

Similarly, neuroscientist Sam Harris argues that accommodation between science and religion is misguided because it is impossible to separate religion from statements about nature, and if these statements are then refuted, this is an unavoidable strike against religion: "The core of science is not a mathematical model; it is intellectual honesty. Every religion is making claims about the way the world is. These are claims about the divine origin of certain books, about the virgin birth of certain people, about the survival of the human personality after death. These claims purport to be about reality" (quoted in G. Johnson 2006a; see also Harris 2004, 2006).

The many critics of this view have argued, however, that although Darwinism—like all materialistic scientific explanations—makes an interventionist God redundant and unnecessary, it does not *require* that God not exist, nor does it falsify the hypothesis that God exists (e.g., Cartmill 1998; Pennock 1999; Scott 2004; Ruse 2001a, 2007). This follows from the distinction between methodological and philosophical naturalism, as defined earlier: use of the first is required by the logic of science, but science does not require the second. God may be (as Laplace famously may or may not have told Napoleon) unnecessary for scientific explanation, but this does not scientifically falsify God's existence.

These critics focus on several problems of the extreme atheistic view. First of all, "there is a simple matter of logic: 'not necessary that X' does not imply 'necessarily not X' or even 'not X'" (Pennock 1999, p. 334). Second, science by definition is not concerned with the supernatural: "Science," says Pennock, "excludes appeal to supernatural entities as a point of method, and thus it is improper to draw directly the atheistic conclusion that God is ontologically unreal from evolution or any other scientific conclusion" (1999, pp. 335–36). Even if it wanted to, science could not test hypotheses about the supernatural. Because "explanations involving the supernatural cannot be tested or falsified, science cannot employ supernatural explanations. Science cannot confirm or deny the existence of the supernatural, or a Creator. Such questions are simply beyond the realm of science" (Kelley 2000). In "any situation, any pattern (or lack of pattern) of data is compatible with the general hypothesis of the existence of a supernatural agent unconstrained by natural law . . . supernatural hypotheses remain immune from disconfirmation" (Pennock 1999, p. 195).

Pennock admits, "it is true that, if someone thought that the biological version of the teleological argument [which "tries to prove the existence of God by saying that God is necessary to explain the apparently designed character of creatures and the fit of organisms to their environments" (1999, p. 333)] was the only reason to believe in the existence of God, then [Darwinian] evolution would indeed be likely to lead that person to atheism" (p. 334). But, as Pennock and many others have noted, almost every theist has more than one reason to believe in God, and "proofs" of God's existence are not usually among the most persuasive (see, e.g., Hasker 1986; Ruse 2007).

Finally, there is an important criticism that is based on neither philosophy nor science. Critics of the extreme atheistic view have observed (correctly, in my view) that in the stridency—even harshness—of their advocacy for secularism, atheists are actually doing more to hurt the cause of science than to help it. There is a fine line between the intellectual honesty of calling a spade a spade and the intellectual bullying of sticking your finger in your opponent's eye; by crossing this line, extreme atheists are giving undecided Americans reasons to move away from, rather than toward, science in general and evolution in particular (e.g., Krauss 2006; Ruse 2007). As Krauss puts it: "Science does not make it impossible to believe in God. We should recognize that fact and live with it and stop being so pompous about it" (quoted in G. Johnson 2006a).

Exactly how widely held such atheistic views are among scientists in general is much more difficult to determine than quoting selected opinions. In the only recent general survey, Larson and Witham (1997) found that around 40 percent of a random sample of 1,000 U.S. scientists from a variety of disciplines believe in a "personal God" (defined as "a God to whom one may pray in expectation of receiving an answer"). This figure was remarkably close to the result obtained in a similar survey eighty years earlier (Leuba 1916). However, in a survey of 255 biological and physical scientists who are members of the U.S. National Academy of Sciences (NAS), Larson and Witham (1998) found that only 7 percent expressed belief in a personal God, and this number was lower than in results of comparable surveys of "greater" scientists in 1914 and 1934. Similarly, in a survey of 151 evolutionary biologists who are members of national academies of science in twenty-two countries, Graffin (2003; Graffin and Provine 2007) found that 80 percent do not believe in a traditional God (defined as "an entity that exists beyond the scope of our observations that is responsible for designing and maintaining life on earth"), and only 5.4 percent do believe in such a God.

Based on these results, it seems that although the great majority of "leading" scientists are effectively atheists, a significant proportion of scientists in general accept some form of supernatural deity. (It would be fascinating to know what proportion of practicing evolutionary biologists, paleontologists, and historical geologists who are *not* NAS members say they believe in a personal God. To my knowledge, no such survey has ever been conducted.) Thus it appears to be the case that as many as two in five American scientists are *not* atheists, and evidently get along quite well professionally in this condition.

Much more is known, of course, about the religious views of the American public. Polls are consistent in reporting that a very large percentage of the American public identify themselves as believers and only a small percentage as atheists. A 2006 Harris poll, for example, found that 88 percent of those surveyed were "absolutely" or "somewhat certain" that God exists, while only 12 percent were "absolutely" or "somewhat certain" that God does not exist.[3] This finding is roughly consistent with those of previous polls. A 2006 CBS News poll found that 82 percent of those surveyed believed in God, while 9 percent believed in "some other universal spirit or higher power," 8 percent believed in neither, and 1 percent were unsure. A 1998 Harris poll and a 2000 *Newsweek* poll both found that 94 percent of American adults surveyed said they believed in God.[4]

Most Americans, furthermore, say they do not hold positive opinions about atheism: "That label [atheist] evokes a strong negative response in American life. . . . It is religion, having a faith, that makes people, in the American context, seem trustworthy, like good citizens and good neighbors. So if you don't have that as a moral boundary, you're an outsider, an other, and, perhaps, a dangerous other" (Edgell 2006; see also Edgell, Gerteis, and Hartmann, 2006).

COMPLETE SUPERNATURALISM: "THE OVERTHROW OF MATERIALISM"

At the extreme other end of the God Spectrum ("A" in figure 10.1), we find creationists of a wide variety of stripes, including advocates for intelligent design. The modern ID movement was founded in the early 1990s, but it found its true legs with the infamous Wedge Document, leaked on the Internet in 1999 (see P. E. Johnson 2000 for a spirited defense of the document, and Forrest and Gross 2004 for a thorough critique; see Discovery Institute 2003 for an authentic copy of the document and an official response to critics). According to the Wedge Document, the movement's goals include "nothing less than the overthrow of materialism and its cultural legacies," and replacing science as currently practiced with "a science consonant with Christian and theistic convictions." Thus, ID "aspires to change the ground rules of science to make room for religion, specifically, beliefs consonant with a particular version of Christianity" (Jones 2005, pp. 28–29).

I cannot help but suspect, however, that (despite their protestations to the contrary; see Discovery Institute 2003) the authors of the Wedge Document either have not really thought their arguments through very carefully or are simply being dishonest, inconsistent, or muddleheaded. How do they, for example, reconcile accepting some aspects of materialistic science (e.g., medicine, agriculture, and engineering), while rejecting others (evolution by natural selection)? The answer appears to be simply that they *don't* reconcile these views. Rather, they pick and choose the science they like—that which does not appear to threaten their worldview—while rejecting the science they perceive as damaging to Western, Christian, socially conservative ideals.

Pennock (1999) has analyzed what one might call this "consistency conundrum" in some detail. He notes, as have others, that modern creationism is attacking not just evolutionary biology and historical geology, but almost every other field of science as well. He writes that "to toss evolutionary theory and scientific naturalism onto the pyre would be to commit much of the rest of science to the flames as well" (Pennock 1999, p. 340), and he expands on what this would mean for practitioners in those other fields as well as for those who use their results.

> Science is godless in the same way that plumbing is godless. Evolutionary biology is no more or less based on a "dogmatic philosophy" of naturalism than are medicine and farming. Why should [ID advocate Phillip] Johnson and his allies find methodological naturalism so pernicious and threatening in one context and not in others? Must we really be seriously "open-minded" about supernatural explanations generally? . . . Surely it is unreasonable to complain of a "priesthood" of plumbers who only consider naturalistic explanations of stopped drains and do not consider the "alternative hypothesis" that the origin of the backed-up toilet was the design of an intervening malicious spirit. Would it not be bizarre to reintroduce theistic explanations in the agricultural sciences and have agronomists tell farmers that their crop failure is simply part of God's curse upon the land because of Adam's disobedience, or suggest that they consider the possibility that the Lord is punishing them for some moral offense and that it might not be fertilizer they need but contrition and repentance? (Pennock 1999, pp. 282–83)

> Consider the medical sciences. It was once commonplace to attribute the origin of certain illnesses to curses or demonic possession. . . . If we accept the intelligent-design creationist's diagnosis, medical schools and research physicians are doing a terrible disservice by not teaching students how to perform exorcisms and by not taking seriously the possible supernatural origins of diseases. (Pennock 1999, p. 283)

> If it is science's naturalistic methodology that is inherently problematic, then Johnson should be equally worried about chemistry and meteorology and electrical engineering. He should also be concerned about automobile me-

chanics, for this field too proceeds under the assumption that God does not intervene in the workings of the motor. But surely no one thinks that these naturalistic sciences imply that God does not exist. (Pennock 1999, p. 333)

Paraphrasing Dennett (1995), Michael Ruse makes the same point: "If we were falsely accused in court, for example, we would be very upset if the judge simply threw out the evidence in our favour and intuited the 'truth.' And we would think you slightly crazy if you went to a surgeon who was guided solely by a little voice from within" (Ruse 2001a, p. 140).

Advocates of ID claim that they are challenging the "philosophy of scientific materialism, not science itself" (Discovery Institute 2003). This is a useful rhetorical approach in their public statements as it allows them to appear moderate and reasonable, and unthreatening to fields such as medicine, engineering, and agriculture that have obvious personal and economic benefits. Yet they never deal with the central problem of this approach: there can be no science in any meaningful sense of the word *without* at least methodological materialism (as defined earlier). Allowing (or compelling) science to include the possibility of nonmaterial, supernatural causes would ultimately destroy science, because resorting to such causes is always possible, fails to spur deeper investigation, and leads to no additional understanding. "Without the binding assumption of uninterruptible natural law there would be absolute chaos in the scientific worldview," says Pennock. "Supernatural explanations undermine the discipline that allows science to make progress." They are too easy and therefore are "the explanation of last resort" and "the poor person's explanations, or rather, the explanations of the intellectually poverty-stricken, since they are available for free" (Pennock 1999, p. 294).

Such predictions of catastrophe might be dismissed as hyperbole, except that the very definition of science was indeed assaulted directly through the political process in 2004–2005 when the Kansas State School Board voted to change the definition of science in the state standards so as to allow for supernatural causes to be included as potentially valid explanations for natural phenomena. In November 2005, the language in the standards (adopted in 2001) was changed from "Science is the human activity of seeking natural explanations for what we observe in the world around us" to "Science is a systematic method of continuing investigation that uses observation, hypothesis testing, measurement, experimentation, logical argument and theory-building to lead to more adequate explanations for natural phenomena" (Overbye 2005). (On February 13, 2007, the Kansas State Board of

Education rejected the 2005 revision, reestablishing science as restricted to the investigation of physical phenomena.)

How do creationists deal with the consistency conundrum? What does Phillip Johnson do when he gets sick or wants to have his car repaired? He appears to want us to believe that he uses science, and the technology that comes from it, just like the rest of us: "The possibility that divine intervention may occur . . . emphatically does not imply that all events are the product of an unpredictable divine whimsy" (P. E. Johnson 1995, p. 92). But, as Pennock points out (1999, p. 298), Johnson never explains what the methods are for determining when divine intervention is and is not to be considered potentially responsible for a particular event. Other creationists, while acknowledging that some theologies may describe God as capriciously and frequently intervening in the natural workings of the world, argue that a correct Christian theism "holds that secondary causality [causation by natural laws] is God's usual mode and primary causality [direct divine acts or miracles] is infrequent, comparatively speaking," and that science should not defer to such primary causality "willy-nilly" (Moreland 1989, p. 226, quoted in Alters and Alters, 2001, p. 151). But, again, there is no definition of "willy-nilly." Like all attempts at "scientific" creationism, ID thus want it both ways: materialistic science when it suits them and supernatural intervention when it doesn't, with no objective rules or regularities to explain or predict why one and not the other in any particular case.

ACCOMMODATION: THE BROAD MIDDLE

Accommodation among Nonscientists

Moving away from the extremes of the spectrum ("B" through "F" in figure 10.1) means admitting the possibility that not everything in the universe is completely and necessarily caused by only material phenomena, and that some kind of supernatural force may be involved, occasionally or continually, either only in the past or continuing into the present. The view that religion and naturalistic science can be accommodated also has a very long history. Ruse (2001a, p. 51), for example, attributes to Augustine (354–430 CE) "the influential thesis that Moses (the supposed author of Genesis) had to write in metaphorical or allegorical form, because the ancient Jews were untutored in science. What Moses said was not false, but not necessarily the literal truth." Similarly, John Calvin's "famous doctrine of accommodation" (as Ruse puts it) "recognized that the Bible is sometimes written in such a form as to make itself intelligible to scientifically untutored folk who would not have followed sophisticated discourse" (2001a, p. 53). In other words, because the works of God as represented by scientific observation of the natural world cannot by definition contradict the word of God represented in scripture, if a literal reading of scripture contradicts scientific conclusions, there is room for interpretation that would render scripture consistent with nature. (William Jennings Bryan made essentially the same concession on the witness stand in the 1925 Scopes trial.)[5]

One of the strongest arguments against the "conflict" model of interaction between science and religion is the historical observation that both Catholic and Protestant Christianity have in a variety of ways been important nurturers and supporters of scientific pursuits at various times over the past thousand years (e.g., Merton 1938; Hooykaas 1972; Cohen 1990; Lindberg 1992; Harrison 1998; Heilbron 1999; Grant 2001). Similarly, despite the flurry of controversy after publication of *On the Origin of Species* in 1859, among "sophisticated churchmen" in Europe and America there was initially considerable accommodation between Darwin and religion (e.g., J. R. Moore 1979; Livingstone 1984; Roberts 1988). The mid- to late nineteenth century was, furthermore, a time in which the accommodation of science and religion was of particular popular interest in both Britain and America (Kloppenberg 1986; Croce 1995). Yet this accommodation was not always especially clearly thought out nor were there abundant calls for its intellectual consistency. As noted by Kazin (2006), in the mid-nineteenth century,

> popular religion [in America] skirted the grand controversy between faith and science. . . . While Darwinism led theologians to either shudder at or guardedly welcome what one historian calls "the breakup of an intellectual system that had endured from the beginning of European civilization," most Protestants seemed comfortable both with a supernatural faith and with the rigors and pleasures of modern America. They prayed every day and went to the theater on Saturday, agreed that one could not serve both God and Mammon yet consumed at least as lavishly as their incomes allowed, fretted about the afterlife and fought for their rights in the present. Bathing their faith in sentimental hues, ordinary Americans ignored what appeared to seminarians and philosophers as glaring contradictions. William James once tried to explain the motivations of religious Americans . . . [when he said]: "As a rule we disbelieve all facts and theories for which we have no use." Outside the lecture hall, few Christians doubted the utility of their own convictions. (p. xviii)

In many respects, this same ambivalent accommodation today represents the mainstream of liberal-to-moderate America. This view is reflected in many areas of public discourse, including most notably the judicial opinions that have successfully kept creationism out of science classrooms in public schools. In his resounding opinion against allowing ID into the Dover, Pennsylvania, public schools, federal judge John E. Jones III, wrote, "Both Defendants and many of the leading proponents of ID make a bedrock assumption which is utterly false. Their presupposition is that evolutionary theory is antithetical to a belief in the existence of a supreme being and to religion in general. Repeatedly in this trial, Plaintiffs' scientific experts testified that the theory of evolution represents good science, is overwhelmingly accepted by the scientific community, and that it in no way conflicts with, nor does it deny, the existence of a divine creator" (2005, p. 136).

Leading advocates for evolution in public schools go to great lengths to say and show that evolution and religion are not incompatible. Richard Dawkins (2006, pp. 66–69) has labeled this approach "the appeasement lobby" and "the Neville Chamberlain school of evolutionists" (see Ruse 2007 for a rebuttal of this label).

> The strength of students' emotional ties to their religious beliefs cannot be overemphasized, and neither can the potential for negative educational outcomes when students feel that their beliefs are in conflict with what they're being taught. . . . The way for science instructors to deal with the issue [of whether evolution and the Bible are in conflict] is by helping students understand that evolutionary science doesn't deny the existence of a supreme being—that evolution simply doesn't address such non-scientific questions. (Alters 2006, pp. 117–18)

> If evolution is presented as antithetical to religion (which is precisely how organizations such as the Institute for Creation Research present it), it is no wonder that a high percentage of Americans reject it. . . . As teachers and scientists, we need to leave an opportunity for the religious individual to work out the accommodation according to his or her beliefs, and not slam the door by inserting extra-scientific philosophical statements about purpose and meaning into our discussions of evolution. (Scott 1996, p. 17)

> Evolution does not necessarily lead to atheism, and if defenders of evolution regularly made this clear it might open the fearful hearts of their audience, which is the first step to opening their minds to the evidence. . . . Defenders of evolution would help their case immeasurably if they would explicitly reject the creationists' contention that evolution is atheistic, and reassure their audience that morality, purpose, and meaning are not lost by accepting the truth of evolution. (Pennock 1999, pp. 336–37)

> If we are able to calm the divisive fears that evolution is the root of an atheistic philosophy that leads to purposelessness and immorality and reassure the creationist that evolution does not bar the roads to meaningfulness, then, and perhaps only then, will the creationist controversy be put behind us so that we can travel those roads together. (Pennock 1999, pp. 339–40)

The value of this argument is highlighted by a story related to me by evolutionary biologist Douglas Futuyma (pers. comm., April 2006):

> Jim Elser, ecosystem ecologist at Arizona State University, teaches a biology course for nonmajors, and for a few years has polled students (after discussing what science is and why "scientific creationism" isn't science) on their view of whether or not scientific creationism should be taught as a scientifically valid alternative to evolution. In the past, an overwhelming proportion of the students say yes, it should be. This year, he gave them an assignment to learn about the position that one of the major religions (assigned at random to a student) takes on evolution. The poll results were exactly opposite to last year's; the great majority of these students say that creationism should NOT be taught! Elser thinks that this is the consequence of learning, by active research, that most religions do not exclude belief in evolution.

Evidence to support this claim of nonconflict comes in the form of numerous religious denominations and leaders who have gone on record as insisting that evolution and religion can be and are readily accommodated with each other. In 1969, for example, the United Presbyterian Church in the U.S.A. set, and in 1982 and 2002 reaffirmed, its official position that "there is no contradiction between an evolutionary theory of human origins and the doctrine of God as Creator." In 1992, the United Church of Christ stated, "We acknowledge modern evolutionary theory as the best present-day scientific explanation of the existence of life on Earth; such a conviction is in no way at odds with our belief in a Creator God." Other major denominational organizations, including the American Jewish Congress, the Central Conference of American Rabbis, the General Convention of the Episcopal Church, the Unitarian-Universalist Association, and the United Methodist Church all have over the past twenty-five years issued statements opposing teaching "scientific creationism" in the public schools. The plaintiffs in the landmark 1981 Arkansas case *McLean v. Arkansas Board of Education,* who successfully sought to overturn the imposition of "balanced treatment" for creationism and evolution in the public school classroom, included Jewish, United Methodist,

Episcopal, Roman Catholic, African Methodist Episcopal, Presbyterian, and even Southern Baptist clergy (see Matsumura 1996 for details).

Despite the uneven relationship of the Roman Catholic Church with science in general, and Darwinism in particular (G. S. Johnson 1998; Artigas, Glick, and Martinez 2006), in 1996 Pope John Paul II proclaimed that the theory of Darwinian evolution is so well supported by so much evidence that it has become "more than just a hypothesis." Evolution, said the pope, is fully compatible with Christian faith and a valid explanation of the development of life on Earth, with only one major exception: the human soul. "If the human body has its origin in living material which preexists it," the pope said, "the spiritual soul is immediately created by God" (1996).

Partially as a spinoff of the increasingly widespread celebration of Darwin Day on Charles Darwin's birthday (February 12) (e.g., Chesworth 2002; Allmon and Grace-Kobas 2007), the Clergy Letter Project began in 2004 as "an endeavor designed to demonstrate that religion and science can be compatible and to elevate the quality of the debate of this issue." As of February 2007 the project's "open letter concerning religion and science" had more than 10,500 signatures. February 12, 2006, was the first Evolution Sunday in more than two hundred churches across the United States, in which ministers devoted their sermons to seeking accommodation between evolution and religion (Clergy Letter Project). "One of the goals of the Clergy Letter Project," says its founder, Michael Zimmerman, a Wisconsin college administrator and biologist, "is to demonstrate that the choice that people are trying to foist on them is a false dichotomy. The fact that thousands of clergy are standing up and saying, 'We are comfortable in our beliefs, in our faith, and in our God, and we are comfortable with modern science,' is a very forceful statement" (Eisenberg 2006).

A similar grassroots effort is the Nebraska Religious Coalition for Science Education, a "network of Nebraskans from diverse religious faiths with the shared conviction that academic freedom, religious freedom, and scientific integrity are indeed compatible." The mission of the coalition is to "proclaim the compatibility of good science (including evolution) and good theology (including creation)," and they furthermore "desire to help raise awareness, particularly in Nebraska's religious communities, that methodological naturalism is an essential part of good science, and that the scientific search for natural explanations is not anti-religious" (Austerberry 2003).

The mainstream (and frequently politically liberal-leaning) media similarly largely embrace accommodationism. *New York Times* science writer William

Broad, for example, writes that "the truth is that science and spirituality, rather than addressing similar ground, speak to very different realms of human experience and, at least in theory, have the potential to coexist in peace, complementing rather than constantly battling each other. . . . The scientists who make sweeping metaphysical claims may represent a vocal minority. But hubris and celebrity are a potent mix, and threaten to intensify a cultural war that need not be" (Broad 2006). The "sermons" of Dennett and Dawkins "are unsatisfying," writes another senior *Times* science writer, Cornelia Dean (2006): "Of course there is no credible scientific challenge to Darwinian evolution as an explanation for the diversity and complexity of life on earth," she continues. "So what? The theory of evolution says nothing about the existence or nonexistence of God." Indeed, the generally negative literary reception to Richard Dawkins's most recent (2006) strident attack on religion (e.g., Bakewell 2006; Holt 2006; Krauss 2006; Orr 2006; Ruse 2007; but see also G. Johnson 2006b and especially Weinberg 2007) suggests that his brand of extreme secularism/materialism is not widely shared among intellectuals.

Accommodationism, however, unquestionably has significant logical problems. First of all, it is simply not always possible. It is sometimes logically impossible to hold specific religious tenets about the world *and* accept current scientific thinking on these same aspects of reality: "Although evolution does not attack religion it does pose the problem of what to do when convincing scientific conclusions come into conflict with the beliefs people hold without evidence" (Bambach 1983, p. 853). "I know it passes in polite company to let people have it both ways, and under most circumstances I wholeheartedly cooperate with this benign arrangement. But we're seriously trying to get at the truth here" (Dennett 1995, p. 154). Furthermore, although Pennock (1999, p. 339) suggests that there is nothing about Darwinian evolution per se that rules out "notions of purpose and meaning," such as that "our purpose in life is to praise God, to accept Jesus as our Lord and Savior, to ask for spiritual grace and . . . to ultimately fulfill our purpose in eternity," atheists such as Provine and Dawkins would object and insist that: (1) it is absurd to praise a God that either does not exist or does nothing of consequence; (2) accepting a person who has been dead for almost two thousand years as a "Lord and Savior" is either meaningless or ridiculous; (3) "spiritual grace" has no meaning at all; and (4) the idea that anything about us besides our atoms will last into eternity is nonsense. Indeed, upon closer inspection, most attempts at accommodation seem to be able to persist only because they lack specificity and logical consistency, and so risk being ultimately unsatisfying to someone who wants to know exactly how God interacts with the physical universe.

General Explanations for Accommodationism

There either is a God or there is not, and therefore either accommodationism is a valid approach to comprehending the material universe or it isn't. Unfortunately, we will never know for sure which is true. So let me first consider why, if it is *not* valid (i.e., God does *not* exist), attempts at accommodation persist, and why most Americans maintain that it is possible to both believe in a supernatural God and accept that naturalistic science works. There would appear to be at least five general potential explanations; certainly more than one could be responsible for the expressed beliefs of a single individual.

Poor understanding of science. Sad though it may be to admit, a major reason that many people can uncritically amalgamate religious and scientific viewpoints may be that they just don't understand science very well (Hazen 2002; Shermer 2002; National Science Board 2006). This poor understanding allows them to believe that the science that, for example, produces the technology they employ and enjoy in their everyday lives is the result of one kind of thinking, whereas ideas such as the Big Bang or evolution by natural selection come from some other kind of much more questionable activity.

Tolerance. Americans have a genuine and famous propensity for tolerance and compromise, especially regarding religion. The "continuing power of the sentiment for harmony" between science and religion is deeply embedded in the American psyche and culture, and this has encouraged a wide and creative range of accommodations (Croce 1995, p. 13). Lacking a homogeneous state-sponsored religion and embracing democratic pluralism as they do, when confronted with an either/or choice, Americans tend to say about most matters, "it depends," "live and let live," or simply "both." That everyone can believe whatever he or she wants seems the essence of the religious freedom guaranteed by the Constitution and it is what most of us seem to want to be able to say.

Hope. One potent force behind widespread accommodationism is clearly the ubiquitous human needs to feel cared for and to have a purpose in life. These needs have long been identified as an important component of religious faith, and are often pointed to as major problems in acceptance of the completely materialistic/secularistic viewpoint. Distinguished evolutionary biologist (and former Dominican priest) Francisco Ayala, for example, describes this as a major issue for evolutionists: "There are six billion people in the world," he says. "If we think we are going to persuade them to live a rational life based on scientific knowledge, we are not only dreaming—it is like believing in the fairy godmother. People need to find meaning and purpose in life. I don't think we want to take that away from them" (G. Johnson 2006a).

These very American and very human feelings also have potentially powerful political implications. Barack Obama, for example, writes in his recent personal manifesto that the political and societal success of evangelicals in the United States "points to a hunger for the product they are selling, a hunger that goes beyond any particular issue or cause."

> Each day, it seems, thousands of Americans are going about their daily rounds . . . and coming to the realization that something is missing. They are deciding that their work, their possessions, their diversions, their sheer busyness are not enough. They want a sense of purpose, a narrative arc to their lives, something that will relieve a chronic loneliness or lift them above the exhausting, relentless toll of daily life. They need an assurance that somebody out there cares about them, is listening to them—that they are not just destined to travel down a long highway toward nothingness. . . . [O]ver the long haul, I think [progressives] make a mistake when we fail to acknowledge the power of faith in the lives of the American people, and so avoid joining a serious debate about how to reconcile faith with our modern, pluralistic democracy. . . . When we abandon the field of religious discourse . . . when we discuss religion only in the negative sense of where and how it should not be practiced, rather than in the positive sense of what it tells us about our obligations toward one another; when we shy away from religious venues and religious broadcasts because we assume that we will be unwelcome—others will fill the vacuum. And those who do are likely to be those with the most insular views of faith, or who cynically use religion to justify partisan ends. (Obama 2006, pp. 202, 213–14)

Social acceptability. People may say that they believe things that they really do not, or aren't sure about, because they think that affirming them is the right or socially desirable thing to do. This "social desirability" or "social acceptability" factor—the tendency to give a favorable picture of oneself—is "generally considered to be a major source of response bias in survey research" (DeMaio 1984, p. 257). For example, "people who feel that being aware of the latest media releases is highly desirable are more than twice as likely to inaccurately report having read nonexistent books or seen nonexistent movies as people who rate such awareness as highly undesirable" (p. 271). The sources of these expectations or values influencing answers can be the person himself, the perception of the interviewer, or society as a whole. People may therefore say they believe in God, when they really do not or are undecided, because they think that this is the socially acceptable answer.

Self-delusion. Psychologists label as "self-delusion" the state of intentional or unintentional psychological dissociation that allows individuals to persuade themselves to believe what they know is not so, or to hold two mutually exclusive beliefs at once (Fingarette 1969; Audi 1988). Psychologists disagree, however, on exactly what causes or allows this phenomenon (see, e.g., McLaughlin and Rorty 1988; Scott-Kakures 1996; Lazar 1999; and references therein). One interpretation is that although the self-deceiver simultaneously holds two incompatible beliefs, one of the beliefs is somehow not consciously "noticed," thus allowing the individual to avoid directly comparing two ideas and so realizing their incompatibility. In other words, we just don't think about it very hard. This simple statement probably applies to a multitude of experiences in the daily lives of every human being. We all hold self-contradictory ideas, "if only because we cannot see far enough into the implications of each of our beliefs" (Fingarette 1969, p. 14). None of us can consciously process everything all the time or be completely consistent in our thinking. Gamblers and other risk-takers, for example, daily go against the odds in the hope of success. We all engage in wishful thinking; we "try not to think about" ideas or conclusions we find unpleasant; and, in the extreme case, we may even completely suppress things that are too painful to have in our consciousness. (See Trivers 1985 for an evolutionary explanation of self-deception.) Such self-delusion or deception is thus not unique to religious people, nor is it necessarily an undesirable characteristic.

Accommodation among Scientists

But what if God *does* exist? If scientists believe this as a starting point, they must in some way reconcile their materialistic pursuit of exclusively natural causes for physical phenomena with their acceptance of an active supernatural deity. The challenge in such reconciliation may be phrased as follows: If God actually intervenes in the lives and events of humans by supernaturally altering the behavior of matter or energy—that is, by performing miracles—with any significant frequency, how can scientists be confident that anything they are studying is not also a miracle? How can we believe in the laws of science if God can and does change the rules? The existence of "lawful regularity" (Pennock 1999, p. 195) is central to the functioning of science, but if it is susceptible to the unpredictable intervention of the supernatural, how can science proceed? How, in other words, how can one be a "religious scientist"?

Attempts to answer this question have a long and illustrious history, including natural theology and Enlightenment deism (see, e.g., Olson 2004; Thomson 2005; Witham 2005). In the mid-twentieth century several prominent evolutionary scientists were explicit and devout Christians, including geneticists R. A. Fisher (1890–1962) and Theodosius Dobzhansky (1900–1975), and mammalian paleontologist (and Jesuit priest) Pierre Teilhard de Chardin (1881–1955). Today, accommodationist scientists include a spectrum of views, from the fringe of "creation scientists" and "intellectuals who find Darwinism unconvincing" (Dembski 2004; see analysis in Chang 2006), to serious scientists who are clearly committed to both serious science and theistic religion and who have clearly thought deeply about exactly how the natural and the supernatural can both be apprehended.

I have been doing a very informal poll of scientists and engineers at Cornell University who have publicly said they are religious, and one shared his beliefs with me:

> I can do science if I believe that the thing I am studying operates according to purely mechanistic principles. I am not required to presume that all of what has happened in the universe for all time is explainable based on purely mechanistic principles. I believe in a God who performs miracles, who hears prayers and sometimes answers them miraculously, and who created the entire physical universe and all life based on principles that cannot be explained by the scientific method. At the same time, I believe that He created the universe in a way that functions according to understandable principles.

When I asked this professor how he could believe in both miracles and materialism, he said that "there is a certain level of trust" between him and God. Basically, he trusts that God will not mislead him by performing a supernatural miracle on something he is working on. He cited the Old Testament for support of this view: "In speaking to the prophet Jeremiah about the certainty of His promise to restore the Jewish people after their impending exile to Babylon, God says, 'If I have not established my covenant with day and night and the fixed laws of heaven and earth, then I will reject the descendants of Jacob and David my servant and will not choose one of his sons to rule over the descendants of Abraham, Isaac, and Jacob. For I will restore their fortunes and have compassion on them'" (Jeremiah 33, 25–26). "Thus," this professor said, "this passage presumes that it is obvious that God has promised to run the universe by fixed physical laws. Thus, the Christian scientist has every right to explore what those laws are." This, he said, "puts God on record as normally not intervening with miracles."

Another Cornell professor told me,

> I believe . . . that our universe was created by God, who sustains it in being moment by moment. The patterns of material behavior that we humans observe, systematize, and call the laws of nature are merely the "customs of God." In other words, God works through the secondary causes, which we scientists study. . . . Methodological naturalism in scientific method, however, does not preclude the possibility of miracles in history. Miracles are precluded neither by philosophical presuppositions (à la Hume) nor by science. They are rare and spiritually significant historical events, the truth of which must be evaluated on the basis of empirical evidence and the reliability of the witnesses.

Somewhat further toward the scientific end of the spectrum are a large (and growing) number of practicing scientists and scientifically trained philosophers and theologians who have recently published book-length expositions of their personal attempts to reconcile religion and science. These include Harvard astronomer Owen Gingerich (2006); physicist and theologian Ian Barbour (1990, 2000); John Polkinghorne, fellow of the Royal Society and particle physicist turned Anglican priest (1998, 2005); cell biologist and Brown professor Kenneth R. Miller (1999), who testified for the plaintiffs at the 2005 Dover ID trial; philosopher of evolution Michael Ruse (2001a), who testified for the plaintiffs in the 1981 Little Rock scientific creationism trial; paleontologists Simon Conway Morris (2003), Stephen Godfrey (Godfrey and Smith 2005), Keith B. Miller (2003), and Daryl Domning (2006); head of the Human Genome Project, Francis Collins (2006); and ecologist Joan Roughgarden (2006).

Three paleontologists and their gods. Three prominent paleontologists who are publicly religious but have not published their own book-length personal religious odysseys agreed to share with me their thoughts on how they reconcile their faith with their science.[6] The three form their own small God Spectrum, and I will consider them here in the order of the more to the less religious.

Peter Dodson. Peter Dodson is a leading dinosaur paleontologist, professor of both geology and veterinary anatomy at the University of Pennsylvania, and coauthor of a major college text in evolutionary biology (see E. O. Dodson and P. Dodson 1985; P. Dodson 1996). He also says he is a devout Catholic theist (P. Dodson 2004). "The idea of an inherent conflict between science and religion," he writes, "is at best a crude caricature, and at worst an outright fraud" (2004, p. 1). "I can accept the insights of evolution by natural selection," says Dodson, "and still have the feeling that this is not the whole story" (1999, p. 190). "The profound intuition of the religious believer," he says, "is that life has meaning and purpose, and this intuition provides the basis for cosmic hope. . . . It is also a profound religious intuition that our existence on this planet is not a matter of chance or accident; we are not the unintended consequence of the uncaring Cosmos" (2006, pp. 25–26). Dodson believes in an ongoing creation: "God's providential love for us is infinite, and by definition this cannot be poured out in an instant but is necessarily on-going and open-ended. Creation is not finished, stars are exploding, comets are impacting, new elements are being created, life is evolving" (1997, p. 8).

Dodson accepts the possibility of at least occasional physical interventions by God in the affairs of the material world, but holds that we will likely not be able to detect them by scientific means. "If divine intervention actually took place by altering the course of evolution by genetic manipulation, for example by non-random mutation, could it in principle be recognized?" (1999, p. 190). "If I have an experience of God, my experience is subjective and therefore inadmissible in science. If a heavenly messenger appears to me once, and my experience is not replicable on demand, this again is excluded, because science is supposed to be repeatable. . . . A religious world view subsumes a scientific view but treats human experience, including contact with the divine, with equal seriousness" (p. 191).

At least some of these interventions, for Dodson, qualify as "miracles." He says, for example, that he accepts "the divinity of Jesus Christ, and most especially his Resurrection" (1999, p. 191)—presumably indicating that he literally believes that a flesh-and-blood human being was stone-cold dead and then came back to life via supernatural intervention by God—and he says he is open to the possibility of nonmaterial medical cures that might be of the sort the Catholic Church occasionally investigates as genuine miracles. He believes that God communicates with him personally, although not in any way he can specifically identify, leading him to know things he would not otherwise know. Although he (perhaps a bit playfully) says of his beloved dinosaurs that, "Like all of His Creation, they gave Him praise" and that "God loved them" (1997, p. 8), he also believes that humans alone, among all species that have ever existed, have a special relationship with God: "My God did not come to save prokaryotes but to save humans" (1999, p. 189).

Nowhere in his published work does Dodson explain in any specific way what he means by terms like *cosmic hope*, *providential love*, or even *creation*. He seems to have intuitive definitions of them, in contrast to terms he uses in his science. In our interview, he said that it would never occur to him to attribute an aspect of a dinosaur skeleton he might be studying to supernatural causes, but he could not clearly explain why he was sure this would not be a potential obstacle to his scientific research. He appears to have a comfortable and largely subconscious inner filter that removes the potentially complicating possibility of supernatural intervention in his own work on dinosaurs, but allows it through in how he attributes causation in human thought and behavior—both his own and others'. When I asked him whether his feelings might be different if he was a neuroscientist rather than a paleontologist, and his day-to-day scientific work was concerned with specific questions of where thoughts, intentions, and meanings come from in the blood, electricity, and neurons of the brain, he said he did not know, but the question did not appear to bother him.

Patricia Kelley. Patricia Kelley is a professor of geology at the University of North Carolina at Wilmington, former president of the Paleontological Society, and a former graduate student of Stephen Jay Gould. She did some of the first important empirical tests of punctuated equilibrium (e.g., Kelley 1983, 1984), became an authority on studying predator-prey coevolution and escalation in the fossil record (e.g., Kelley and Hansen 1993, 2003), and has written and spoken frequently against creationism. She also takes great pride in being married to an ordained Presbyterian minister and in having taught Sunday school—specifically adult Bible study class—for most of the past thirty years (see chapter 9; Kelley 2000, 2009). She describes herself as a paleontologist with a deep religious commitment. She agrees that supernatural explanations act as "science stoppers" (Ruse 2001a): in other words, turning to a supernatural hypothesis when natural processes appear to provide insufficient explanations prevents further pursuit of natural explanations. But she also is very concerned that "bright young minds are being forced to choose between faith and science" (see chapter 9), so she has long sought to explain her personal accommodation of science and religion to her students.

Kelley views evolution as God's means of creating; she thinks creation is an ongoing process spanning billions of years and that it is not over yet (see chapter 9). She thinks that God acts through natural processes, but (similar to Dodson) says that even if God did act through supernatural processes, "we couldn't tell it anyway." She says, for example, that she believes in a God "who has the power to intervene in the physical world, but if such intervention occurs it would not be possible to confirm it by scientific testing." She believes in a God who has the power to intervene in any way he wants, including via communication with human consciousness or through physical phenomena, but that hypotheses that invoke the supernatural are untestable and therefore cannot be employed in science. Such hypotheses cannot be disproved, but they are not very useful in understanding the world. She says that "prayer may result in nonmaterial changes (e.g., in human attitudes, desires, perceptions, feelings)," but does not make clear whether this is because God somehow communicates with the human who does the praying or because the act of prayer itself changes the person's attitudes or actions. This vagueness she attributes to her inability to differentiate between the two: "a change in attitude," she says, "could be considered the way in which God is communicating" and the attribution of this change is a matter of faith, not empirical test.

On the written questionnaire, she locates her views between B and C along the God Spectrum (figure 10.1), clarifying her belief as: "A supernatural (nonmaterial) God designed and created the entire material Universe and has maintained and guided it through all time through natural processes that are in accord with known physical laws (e.g., quantum indeterminacy)." She rejects the occurrence of miracles in the sense of interventions by God in the physical world through alteration or suspension of physical laws, and believes that God's interventions are accomplished literally by the action of these laws.

Richard Bambach is one of the leading American paleobiologists of the late twentieth century (see A. I. Miller 2004). He spent most of his career at Virginia Tech and then retired, first to Harvard and then to the Smithsonian Institution, where he is now a research associate in the National Museum of Natural History. He has been an active church member (in either the United Church of Christ or the Presbyterian Church) since his early teens. Bambach does not believe in miracles, "that 'supernatural' events occur," or that "'God' interferes in any supernatural fashion with the universe." He also does not believe in "a God who is personally acting" on his behalf, or that "God acts directly on day-to-day issues that people are capable of dealing with themselves." He does, however, consider himself a "religious person," and he believes that "there is a non-material entity that embodies the qualities and concepts [of] decency, morality, fellowship, stewardship, goodness, helpfulness, caring, [and] love." Bambach holds that "both tenets of religion and widely accepted scientific theories are *beliefs*" (1983, p. 851; emphasis added) and that "we need more than science" to lead a meaningful life.

His views on what that "more" might be are centered on the idea that we are unlikely to be able to understand any deity that may exist with the tools available to us as "limited mortal humans." It is not, he thinks, reasonable to expect "complete and definitive, reproducible and documentable evidence of a physical sort" for what may be "senses or feelings or beliefs" which may "relate to some actual reality, but a reality that is veiled" because it "is immaterial in scientific terms and not fully comprehensible to us."

Even though God is "unknowable and incomprehensible," Bambach nevertheless recognizes God as "the ultimate essence of the range of spiritual values . . . that can inform us of the best way to maximize the quality of life for all living things, which I believe should govern human behavior." Even clearly talking about such things may be difficult, he says, because they "may be beyond our abilities to express." The human inability to comprehend God, Bambach suggests, might be "analogous to how placozoans, sponges, or cnidarians might (or might not) conceive of the intellectual and physical capabilities of humans." (This is reminiscent of Darwin's rumination in his famous 1860 letter to Asa Gray [1993, p. 224]: "With respect to the theological view of the question . . . I feel most deeply that the whole subject is too profound for the human intellect. A dog might as well speculate on the mind of Newton. Let each man hope and believe what he can.")

Bambach's views include some measure of trust that he will not be misled by such a God. Most Christians, he says, "believe (have faith) that God is not deceptive or cruelly intentioned" and that "studying and learning about that physical reality IS learning about what God actually did." Accordingly, he says, "the universe I observe as a scientist IS the physical reality in which the spiritual truths and meanings that I have faith exist (and that are the root of my religious orientation) operate—or can be applied." He too has "faith that the universe and all in it that I can observe are what should exist and that whatever supernatural entity may relate these things together is not 'jerking me around.' "

As a way of further exploring this idea of the existence of the supernatural, Bambach says that he can "think of two things that can embody a very active concept of a supernatural entity that, in principle, could exist" and not be inconsistent with the conclusions and methods of science:

1. It is possible, he says, "that a supernatural entity created the universe and the laws of nature" in the singularity of the Big Bang, and that this

"designed system could operate to bring to fruition the creative potential formed in that system in its origins so that it would evolve as it has." He explains that he means by this not "a directed existence leading up the ladder to man," but rather "a system in which life could have evolved in many places and times, with some living systems occasionally evolving sentient, intelligent life forms with capabilities similar to (or even superior to) humans'. These sentient systems could begin to conceive of the supernatural entity."

2. It is also possible, he admits with some hesitation, "that the entire universe was created by a supernatural entity just a second ago (or at whatever moment you would like, a second before you read this or a second before I wrote it, or a second before Shakespeare started writing *Hamlet*; it doesn't matter—the principle is the same). If this were done so that everything was in place at the moment of creation, with the energy (light and other radiation) distributed in space as it is (or would have been at the moment of creation so that it would then shift to be distributed as it is now) and with all sentient organisms with the memories they have, and so forth and so on, then that universe would be absolutely indistinguishable from one that has the history we think we are unraveling through scientific study. Interestingly, although entirely supernatural in origin, such a universe would be scientifically studyable so that we could continue to advance technology, since the laws of nature . . . therefore the practice of science and engineering, would be identical to those in a universe with the natural long history we believe it has had."

Either of these two scenarios, says Bambach, could have happened, but since they are "indistinguishable from the universe as we think it is," they "add nothing to our scientific understanding" and so can be ignored as scientifically useless. But, he suggests, "if you want to be honest about what science can and cannot do about supernatural entities," these two scenarios point out that science "is built on the assumption that what we see is the only story that could have happened. And that just ain't so." The "bottom line" says Bambach, is that "everybody has a faith" and that "none of us KNOWS what the 'truth' is. If we did, there would be nothing to discuss."

The nature of modern scientific accommodationism. The three paleontologists with whom I talked and the recent authors of book-length expositions are all intellectually honest persons for whom religious faith provides meaning to their lives and no threat to their science. Their views of how their science and their faith interact differ, however, on just about every other detail. Some accept physical miracles; some do not. Some think that at least some aspects of God or God's actions are scientifically knowable; some do not. Some are comfortable in institutional faiths; others are not. None have worked out what could be called a thorough, detailed, and consistent view of exactly what God is and how we (scientist or nonscientist) can know anything very detailed about it. Some say that even if they could envision—or wanted to envision—such a view, they could not find words to describe it.

There are other authors, however, who have suggested more cynical interpretations of these accommodationist views. One reason might be that such books sell: "Publishers have come to learn that there is a lot of money in God," says Neil Degrasse Tyson, "especially when the author is a scientist and when the book title includes a direct juxtaposition of scientific and religious themes" (Tyson 2001). Will Provine has suggested that accommodationist scientists are driven by a mixture of unconscious pragmatism and outright intellectual dishonesty: "In the United States, elected members of Congress all proclaim to be religious; many scientists [therefore] believe that funding for science might suffer if the atheistic implications of modern science were widely understood" (Provine 1987, p. 52). He later expanded this suggestion:

> I suspect there is a lot of intellectual dishonesty on this issue. Consider the following fantasy: the National Academy of Sciences publishes a position paper on science and religion stating that modern science leads directly to atheism. What would happen to its funding? To any federal funding of science? Every member of the Congress of the United States of America, even the two current members who are unaffiliated with any organized religion, profess[es] to be deeply religious. I suspect that scientific leaders tread very warily on the issue of the religious implications of science for fear of jeopardizing the funding for scientific research. And I think that many scientists feel some sympathy with the need for moral education and recognize the role that religion plays in this endeavor. These rationalizations are politic but intellectually dishonest. (Provine 1988b, p. 69)

It is possible that the scientists seeking accommodation between their science and their religion are victims (or perpetrators) of one of these (or other) factors—poor understanding of science, inordinate tolerance, hope/wishful thinking, social acceptability bias, self-delusion, popularity, cynical pragmatism, or intellectual dishonesty—that render accommodationism a fool's errand. In other words, it is possible that they all are misled. Certainly this is the interpretation that critics of accommodationism, such as Dawkins, Dennnett, and Provine, favor. It is also possible that these accommodationist scientists are struggling toward a truth that is still only poorly defined and certainly difficult to comprehend.

Which is it? As far as scientists are concerned, is there a God or isn't there? How do we figure this out? How do we know if we can? Are attempts to seek accommodation between religion and science legitimate or foolish?

What scientists accept as true about the natural world is not decided by a vote of opinions; it is supposed to be decided by agreement with observation about nature. In practice, however, what is accepted as true in a particular area by the overall scientific community at any given moment is usually the majority view among those scientists who are specialists in that area. If most specialists who have devoted years to researching a topic accept a particular theory, it is usually treated as true by nonspecialists who have not studied it in as great detail; it is what "most experts say." If a view accepted by a majority of specialists is challenged by new information or interpretations, it will generally not be discarded—nor the new view accepted—until enough contrary theory or examples have been put forward to convince a majority of specialists to change their minds. If this doesn't happen, the old theory stands. If there is no clear majority view, then textbooks and "received wisdom" will normally report that the topic or question at issue is "poorly understood" or "controversial."

By this standard, the "scientific status" of accommodationism can be labeled only as unclear. It is not the majority hypothesis of "leading" evolutionary biologists (e.g., members of the NAS), yet it may well be widely accepted by the much larger group of "ordinary" evolutionists and other scientists. Whoever or however many scientists accept it, it certainly cannot fairly be called a single, well-developed, or coherent idea. Indeed, it is entirely possible that the majority of its adherents—even the ones who have thought a great deal about it—cannot articulate it clearly at all. Taking the variety of views about accommodation at face value suggests similarity to a field of science just prior to a paradigm-shifting "revolution" (Kuhn 1970). By the standards by which any other major idea in science—such as plate tectonics, atomic theory, relativity, or natural selection—is normally assessed, the nature or existence of God is thus anything but decided. This, to me, means two things: first, that it is incorrect to state that a consensus exists one way or the other and, second, that ample opportunities exist for new ideas and new interpretations.

The God of Accommodation

If accommodation between religion and science is possible (i.e., God does exist), what is the nature of the God with which such religion would concern itself? Accommodationists—scientists and nonscientists alike—are looking for the existence of an "interested" God, not an abstract or impersonal one (Weinberg 1993 p. 245; Haught 2006, p. 697). Accommodationists, in other words, are "looking to science for something far more specific—the constant, hovering presence of the kind of God described in Sunday school, who watches over us and answers our prayers. This is not the God of deism, who cranked up the universe and let it run" (G. Johnson 2005).

The serious scientists who are trying to accommodate their religion and their science appear to be focusing on three different approaches to finding this "interested" God. These three approaches are usually, but not always, clearly distinguished.

1. God does (or did) perform physical miracles which are outside the realm of normal physical laws, but extremely rarely and only under the most exceptional of circumstances, and so such supernatural interventions can be safely discounted as complicating factors in doing normal science.
2. God acts via existing physical laws and processes, in ways that are specifiable but probably undetectable. For example, Kenneth R. Miller eventually concludes his lengthy discussion of why religion and evolution are compatible with the conclusion that God probably acts through quantum indeterminacy. This idea has been considered favorably by some (Russell 1998, 2006) and criticized by others (Ruse 2001a).
3. God acts via a set of processes that are neither physically specifiable nor detectable, and that are completely beyond the physical reality that is the purview of science and accessible only to the tools of faith. Such a God actively upholds the universe in a way we have not discovered and will not ever discover. And this God communicates with human beings.

In exploring this third approach, both ontology and language are significant problems. Science deals in concrete physical ideas such as observation, hypothesis, and theory, whereas religion (especially Christianity) is often said to deal with much different kinds of concepts, expressed in phrases like "the eternal divine mystery" (Haught 2006, p. 698). For example, statements such as the following are common among theologians, but make no sense at all in a scientific context:

"a humble, kenotic, self-emptying God . . . undergoes a *kenosis* (that is, a pouring out) of the divine substance . . . manifested in the obedience and self-sacrifice of Christ. . . . [As John Paul states, the kenosis is] 'a grand and mysterious truth for the human mind, which finds it inconceivable that suffering and death can express a love which gives itself and seeks nothing in return' " (Haught 2006, p. 699).

> God is to be thought of as one who not only creates, but also makes and faithfully keeps promises, inviting (even commanding) people to hope. This God opens up the future even where there appear to be only dead ends. The opening of the future, however, is not to be thought of as an ad hoc, interventionist, readjustment of the laws of nature. Rather, it is a constant aspect of the way God relates to the world. God, therefore, may rightly be called the "power of the future." . . . [T]he revelatory image associated with the picture of Jesus as the manifestation of God is one in which the divine mystery gives itself away to the creation in humble and selfless love. (Haught 2006, p. 704)

Can such religious ideas and language be put into any kind of scientifically recognizable framework (even if not scientifically studyable)? I believe that they can.

What if this concept of "mystery" actually refers, not to some vague and fuzzy sense of an all-powerful but benign force in the universe, but to an actual separate reality, outside the rules and reach of scientific materialistic rational investigation? What if there really is a God, but this God exists and functions in ways that are literally undetectable to science because they function in some parallel realm that interacts with our material universe only in nonmaterial ways? Specifically, such a God might not physically intervene in the material world, either by changing or suspending physical laws, but might communicate with people through their dreams or thoughts, not by physically rearranging neuronal signals but in some other way that we cannot and will not discover by physical investigation. Such a God might therefore be said to "answer prayers," perhaps along the lines of the statement attributed to Unitarian minister Lon Ray Call (1894–1985): "Prayer does not change things; it changes people; and people change things." Humans may be special in the eyes of such a God, but our special qualities, whatever they are, will remain undetectable by any techniques of material science. Perhaps such a God is knowable in some way analogous to science, via the techniques of prayer, meditation, or systematic theology. Perhaps not.

Extreme atheists might argue that such a God is either meaningless or just more wishful theological thinking. Provine, for example, says that "a widespread theological view now exists saying that God started off the world, props it up and works through laws of nature, very subtly, so subtly that its action is undetectable. But that kind of God is effectively no different to my mind than atheism . . . [because] this kind of God does nothing outside of the laws of nature, gives us no immortality, no foundation for morals, or any of the things that we want from a God and from religion" (1988b, p. 70). But if such a God did communicate with and influence humans in some undetectable but significant way, then such a God might affect morals, ethics, and behavior, and so might matter a lot.

As noted by Pennock (1999, p. 335), the idea that an omnipotent, supernatural God could physically intervene at any or all points in the process, and do so in a way that we could never discover scientifically, is an old one. It has its extreme versions, such as "the appearance of age" creationist view, and its less extreme versions, such as God creating the genetic variations upon which natural selection works, but in so subtle a manner that they appear random to us, or God working through quantum indeterminacy (K. R. Miller 1999). All of these conceptions, however, have God working through the existing physical laws and processes. It thus remains possible that naturalistic science could someday discover a way to detect these effects, and they would then no longer be supernatural. If, however, God actually exists, but does so at a completely nonphysical level, then that God is truly consistent with physical naturalistic science.

This conception of God is similar to that of ontological naturalists such as Thomas Hobbes and Baruch Spinoza, who allowed for the existence of God "provided God's attributes are appropriately constrained to conform to the regimen of the given natural ontology" (Pennock 1999, p. 190). It is also similar to the concept of God implied by at least some of the official doctrine of the Roman Catholic Church, which maintains that "according to the Catholic understanding of divine causality, true contingency in the created order is not incompatible with a purposeful divine providence. Divine causality and created causality radically differ in kind and not only in degree. Thus, even the outcome of a truly contingent natural process can nonetheless fall within God's providential plan for creation" (International Theological Commission 2004). Such a God, of course, is a literal "other" as far as the scientific approach to the world is concerned, and so is likely to be extremely difficult for scientists to seriously consider:

> There is no way . . . to understand this paradoxical divine mode of creativity by looking for clear analogies in the thought-world of science with its emphasis on efficient causation. Therefore, attempts to render divine action

intelligible on the analogy of efficient causation or of what passes as causal in the natural sciences will, in my opinion, end up diminishing or obscuring the multifaceted way of influencing the world that theology must attribute to God. For that reason, theology must not apologize for its perpetual failure to arrive at complete intellectual clarity with respect to divine action and divine providence. Theology does not do justice to the power of divine action unless it employs a variety of images that cannot be smoothly mapped onto one another. . . . Theology must avoid exclusive fixation on any of the metaphors it uses. . . . Theology need not be embarrassed that its subject-matter, especially the religious sense of divine action, must always be approached by a tentative and dialectical discourse that constantly allows itself to be relativized by appeal to a rich variety of symbols, analogies, and metaphors. It is a mark of eminence, not the absence, of the divine that we need to clothe our fragile and finite words about it in a rich plurality of references. (Haught 2006, p. 704)

> God does not have to become one actor among others in the cosmic and evolutionary story in order to be profoundly effective. Rather, divine action with respect to evolution and cosmic process is to ground and sustain the narrative loom upon which an indeterminate and still unfinished cosmic drama may be woven. . . . God's grounding and sustaining of the narrative structure of natural being is a much deeper kind of involvement in the world than could ever be the case if divine action consisted essentially of engineering things in the immediate manner that evolutionary materialists, creationists, and intelligent design advocates consider most appropriate to a masterful God. (Haught 2006, p. 707)

God may simply not be understandable at all:

> For my thoughts are not your thoughts, neither
> are your ways my ways, saith the Lord.
> For as the heavens are higher than the earth, so
> are my ways higher than your ways and my
> thoughts higher than your thoughts.
>
> (ISAIAH 55:8–9)

This is not a "God of the gaps" that acts only in those areas of temporary scientific ignorance (Bonhoeffer 1972; Ruse 2001a, p. 122), and so is susceptible to shrinkage with expanding scientific knowledge. This is a "real," personal, meaningful God. And whether or not we personally believe such a God actually exists, this explication means that evolutionists or geologists could honestly describe such a God as not inconsistent with the methods and findings of science; indeed, one might even be able to say that such a concept represents as much as science can say about what God is.

Paleontologist George Gaylord Simpson suggested a similar concept of God in his autobiography, in a chapter entitled "God and I":

> Much has been explained and much more is certainly explicable in terms of the material characteristics or properties of the universe. That does not explain how it happens that the universe has those characteristics or properties, how, indeed, it happens that the universe exists. I, at least, cannot even imagine any possible facts, any conceivable observations that would lead toward such an explanation. That Mysterious Ultimate is, then, inaccessible to scientific, which is to say to rational, human investigation. As far as I am concerned, it is God, or better, god. This god is in full literality ineffable, which means incapable of being expressed, unutterable, indescribable. (1978, p. 29)

I am not saying that I believe such a "nonmaterial God" actually does exist. (Personally, I have no evidence for it and so cannot say one way or the other, although I find such a concept unnecessary to understand the world and hence unlikely.) Nor am I suggesting that atheistic or agnostic scientists should or will accept that it does. I am saying only that such a God *could* exist, and could "answer prayers" and "intervene" in the world in meaningful ways, without interfering in or conflicting with how science goes about understanding the physical universe.

Such a God would clearly not be acceptable to many religious people, especially those who insist on a deity who does intervene at a material level, that is, who causes miracles. Acceptance of a nonmaterial God would inevitably require considerable adjustment on the part of individuals who believe in, for example, the virgin birth and the physical resurrection of Jesus, or the physical intervention of Allah in daily physical events. Yet it is a God that could significantly affect the lives of human beings.

POLITICS, PEDAGOGY, AND REALITY: A SUGGESTED WAY FORWARD

For those who teach about historical geology and evolution, I believe that the previous discussion allows us to reach four important conclusions:

1. *It is bad politics, bad marketing, and simply incorrect to say, "It's either Darwin or God, not both."* It all depends on what one means by "God." The term and concept can have many meanings, and not all of them can be falsified by science. There also *are* many smart, practicing scientists who say they "believe in God," so clearly it is not impossible or irrational to do so. Furthermore, it is simple sociopolitical realism and good pedagogy to attempt to appeal to people from where they are. Telling the average class of American college students that they can have either God or evolution—when the majority of them are probably religious to a greater or lesser degree, and we as scientists are not ourselves speaking with one voice on the subject—is not being completely honest, and it is surely politically foolish. As Peter Dodson puts it, "When forced by scientists to choose between a religion that enriches human experience and an evolutionary science that ignores human experience and minimizes humans as a species, people will unhesitatingly choose the religion that gives meaning to their daily struggles" (1999, p. 192). Barack Obama makes exactly the same point in a political context: "To begin with, it's bad politics [to avoid or denigrate religion]. There are a whole lot of religious people in America, including the majority of Democrats" (Obama 2006, p. 214).

   Atheistic critics of accommodation (e.g., Dawkins 2006; Harris 2006; Hitchens 2007) have little patience for this argument; they think it is hypocritical, misleading, and intellectually spineless and irresponsible. Despite suggestions to the contrary (e.g., Nielsen 1993; Stenger 2007), however, science cannot disprove the existence of the supernatural. We simply have not proved the nonexistence of all Gods, nor disproved the existence of all meaningful supernatural God-like phenomena. This being the case, evolutionists should not allow these extremists—however amusing, popular, or articulate they may be—to speak for all of us (Ruse 2007).

2. *The politically correct, "Let's all get along" view isn't defensible.* Some conceptions of God are simply not compatible or accommodatable with modern science, and that may include the conceptions of many Americans. Those religious traditions that require belief in specific statements about the material world that science has clearly falsified (or at least rendered extremely unlikely) are not consistent with science, and it is

irresponsible to say or imply that they are, or to duck the subject entirely, allowing students to think that they are. Such statements include (but are not limited to) that the Earth is six thousand years old; that all species of organisms were created as we see them today and did not descend from other species; that humans are not descended from other nonhuman species of primates; that evolution has been guided by a supernatural purpose or force, outside of the nature of organisms themselves or their environments. Those who accept any of these ideas as true or reliable statements about the way the material world works are not following the rules of science, not thinking scientifically, and not really understanding or using science at all. Every teacher has as much obligation to refute these statements as to assume that students will accept that matter is made of atoms, that the continents move, and that the Earth goes around the sun. (I would also say that every intelligent layperson reading this has a personal obligation to reach a similar conclusion in the interest of intellectual consistency.)

Furthermore, it seems to me extremely difficult (and perhaps especially problematic for some self-professed religious scientists) to accept a scientific view that allows physical miracles, of any sort, ever, because there is no nonarbitrary criterion for deciding when such supernatural interventions are allowable and when they are not. Such a lack of internal consistency means that some (but not all) attempts to reconcile and accommodate traditional religions, especially Christianity, with modern science are probably not intellectually valid.

3. *Most scientists have not articulated their own personal views.* They consequently find it very difficult to explain those views when asked (as most of us are). Evolutionary biologists and historical geologists should, therefore, as a matter of professional preparation, articulate their own personal theological view, and be ready and willing to explain it, with as much clarity as they can explain natural selection, to students or anyone else who asks. Some of my colleagues may object that this is "too personal" but I simply don't think we have that luxury. We should all write out our own versions of "This I believe" (Allison et al. 2006). It may well be, as previously discussed, that we as humans lack the investigatory and/or explanatory tools to fully explain what we think we know, but we should try our best anyway.

4. *There is always the possibility of things existing that we don't know about,* but this is not the same as telling science students that they can just accept any old thing they want. Consistency is important; without it science cannot function. But, as Emerson (1841) memorably noted, "a foolish consistency is the hobgoblin of little minds." Modern materialistic science is *a* way of knowing about the world (J. A. Moore 1993), not the only way. As discussed above, there *are* rational hypotheses of the supernatural (which could be referred to as God) that are compatible with modern science in that they exclude physical miracles and call on a separate existence that is not, and never will be, accessible to the methods of science. One need not accept (or believe) these hypotheses (I personally do not) to state honestly that they exist—that is, that they *are* logically possible, that they are intellectually consistent, and that they fulfill many (although certainly not all) of the requirements that many religious people have for a "meaningful" God. An essential part of such concepts is that existence may be more than the material universe that is accessible to science. Science must minimize its arrogance and insistence that the material world is all there is if we are to convince nonscientists that we deserve their respect, understanding, and support.

CONCLUDING THOUGHTS

The relationship between science and religion is a difficult topic for many reasons, including that its investigation requires scientists to be self-reflective, which is not easy for many of us. A secular scientist who read an earlier draft of this chapter commented that the four conclusions previously listed are "painfully self-evident." Perhaps they are, but probably only to someone who thinks frequently and deeply about the relationship between science and religion. My experience, however, is that most scientists don't, and that if these conclusions *were* self-evident more scientists would publicly state them. A religious scientist found the same draft "way too sympathetic to the Dawkins/Dennett camp" and suggested that I had made religious scientists out to be "some kind of three-headed monsters that can't be understood by normal people," instead of "persons for whom faith provides meaning" to their lives and no threat to their science. An atheist reader, on the other hand, thought I was much too easy on the religious scientists I talked to and did not press them hard enough to explain the apparent contradictions in their beliefs. These and other comments suggest that I have succeeded in at least sighting, if not reaching, the middle ground where I think a more adequate approach to this topic must lie.

Why is it worth the trouble for scientists to wrestle with such a thorny debate, when we have so many other issues on our daily research and teaching agendas? Why can't we just have both religion and science, let everyone believe what they want, and leave it at that?

It is easy to fall into hyperbole in such an emotional issue, but it is hard to avoid the conclusion that the stakes here are *very* high. Intellectually, evolution is *the* fundamental idea of modern biology. Essentially every field of biology is to a greater or lesser degree based on it. But evolution is about far more than biology. Evolution is also central to many areas of the Earth sciences, and the assumptions and conclusions from other areas of science that underlie evolution—such as the great age of the universe, the solar system, and the Earth; the continuity of past and present processes; and the constancy of physical law in time and space—are shared with other fields of science, such as astronomy, physics, and chemistry. Every major organization of professional scientists in the United States has endorsed the teaching of evolution. Darwinism—evolution driven mainly (but not exclusively) by natural selection—has been the dominant causal hypothesis among biologists for evolution for more than half a century and among the most successful ideas in all of science in its ability to predict and explain observations. Even its scientific critics (e.g., Gould 2002) do not reject its fundamental aspects or evidentiary basis.

Thus, if Darwinian evolution is not valid, then there are likely serious problems with the way we think about science, and with many—perhaps all—other scientific ideas that were generated with the same methods and in which we have equal confidence. Since science has obviously allowed us to achieve an enormous amount of understanding and control of ourselves and our environment over the past four hundred years, this big a mistake would appear to call into question the very nature of our ability to reason in the world.

Broad understanding of evolution is therefore vital to the future of science in society. Clearly, however, the way scientists and educators have been approaching building that understanding is not working. Religion certainly is not the only reason for this failure, but it is surely a major part of the problem. We need a different, or at least an additional, approach to reconciling religion and science, because if left alone, the current controversy between them in the United States threatens to undo much of the social compact that has allowed American science to accomplish so much. The challenge to evolution is just one piece of a wider social/political assault on objective science (e.g., Mooney 2005; Taverne 2005) and, as Kenneth R. Miller has put it, "American science will face a peril of the first order if it fails to understand and to respond effectively to this challenge" (2005, p. 13). It is simply not acceptable for professional scientists to hide in their labs and offices and hope that those we perceive as crazies will go away. They won't. As Michael Ruse says, "If we do not like what the churches are feeding people, we had better come up with an attractive alternative" (2007, p. 38), and this means starting from the viewpoint of where the majority of people are. "For most Americans," as Kenneth Miller notes, "'What about God?' is indeed the most important question" (2005, p. 14).

More broadly, we live in a time of both ubiquitous technological and scientific influence and surging religious fundamentalism, in the United States and abroad. Science and technology are major drivers of the modern global economy, and many of the major challenges—environmental, medical, agricultural—that confront modern society will require scientific and technological solutions in the very near future. Yet the status and future of science education, and even science itself, in the United States and elsewhere, are dangerously uncertain. Fundamentalist religious challenges to science are not only coming from Christians; other fundamentalist faiths, especially Islam, appear increasingly opposed to modern science in general, and to evolution and its associated fields in particular (e.g., Edis 2003; Guiderdoni 2003; Nasr 2006; Weinberg 2007).

The relationship of science and religion is thus not a topic that should be limited to sermons or philosophy classes or books for relatively narrow audiences (such as this one). It should be everywhere, because it is perhaps the single most important problem facing humanity today.

ACKNOWLEDGMENTS

I am enormously grateful to Richard Bambach, Peter Dodson, Robert Fay, Patricia Kelley, Arthur Lembo, and Mark Psiaki for sharing with me their personal views on science and religion; to Doug Futuyma for discussion; and especially to Peter Dodson, George Gorman, Karl Johnson, Tricia Kelley, Kevin Padian, Will Provine, Janet Shortall, Bob Smith, and the students in the Evolution and Religion seminar, which I cotaught with Will Provine at Cornell University during the fall of 2007, for their comments on earlier drafts of the manuscript. Thanks also to Emily CoBabe for inviting me to give a talk at the University of Colorado Astrobiology Institute in 2006, at which some of these ideas were first publicly presented; to Jill Schneiderman for motivating me to write it all down; and to Andrea Kreuzer for assistance with the references.

NOTES

1. I mean this choice of word to be neutral and not at all pejorative. It has been used in a similar sense by, among others, Hooykaas (1972) and Conser (1993). Other possible alternatives include *compatibilist*, *engagement*, *connectivist*, *harmonist*, *conciliationist*, *theistically scientific*, and *theistically evolutionary*.
2. The relationship of Judaism, Christianity, and Islam to science in general, and their reaction to evolution in particular, have varied. Much of what their reactions have had in common, however, relates to their sharing at least the basic aspects encompassed in this definition. In the remainder of this essay, largely for reasons of space, my comments will refer mostly to Christianity, and mainly in the United States, but this does not imply that they do not apply to Judaism or Islam.
3. The 2006 Harris poll also states that "while most U.S. adults believe in God, only 58 percent are 'absolutely certain.'"
4. The 1998 Harris poll and the 2000 *Newsweek* poll also state that a large majority of people believe they will go to heaven, and only one in fifty thinks he or she will go to hell.
5. In trying to answer Clarence Darrow's questioning about the divine inspiration of the Bible, Bryan testified that the Bible "was inspired by the Almighty, and He may have used language that could be understood at that time" (Scopes 1925, p. 286).
6. I interviewed Peter Dodson on May 25, 2006, and received written responses to questions from him on January 19, 2007. I received written responses to questions from Patricia Kelley on January 21, 2007. I interviewed Richard Bambach on October 21, 2006, and received written responses to questions from him on January 28 and 30 and February 4, 2007. Unless otherwise noted, all quotations and paraphrases are taken from these personal communications.

REFERENCES

Abernethy, G. L., and T. A. Langford, eds. 1968. *Philosophy of Religion: A Book of Readings*. Macmillan, New York.

Allison, J., D. Gediman, J. Gregory, and V. Merrick, eds. 2006. *This I Believe*. Henry Holt, New York.

Allmon, W. D. 2009. The Structure of Gould: Happenstance, Humanism, History, and the Unity of His View of Life. Pp. 3–68 in W. Allmon, P. Kelley, and R. Ross, eds., *Stephen Jay Gould: Reflections on His View of Life*. Oxford University Press, New York.

Allmon, W. D., and L. Grace-Kobas, eds. 2007. *Darwin@Cornell 2007*. Special pub. 30. Paleontological Research Institution, Ithaca, New York.

Alters, B. 2006. Evolution in the Classroom. Pp. 105–29 in E. C. Scott and G. Branch, eds., *Not in Our Classrooms: Why Intelligent Design Is Wrong for Our Schools*. Beacon, Boston, MA.

Alters, B. J., and S. M. Alters. 2001. *Defending Evolution. A Guide to the Creation/Evolution Controversy*. Jones and Bartlett, Boston, MA.

Artigas, M., T. F. Glick, and R. A. Martinez. 2006. *Negotiating Darwin: The Vatican Confronts Evolution, 1877–1902*. Johns Hopkins University Press, Baltimore, MD.

Atkins, P. 2006. Atheism and Science. Pp. 124–36 in P. Clayton and Z. Simpson, eds., *The Oxford Handbook of Religion and Science*. Oxford University Press, Oxford, UK.

Audi, R. 1988. Self-Deception, Rationalization, and Reasons for Acting. Pp. 92–120 in B. P. McLaughlin and A. O. Rorty, eds., *Perspectives on Self-Deception*. University of California Press, Berkeley.

Austerberry, C. 2003. The Nebraska Religious Coalition for Science Education. Presentation to the Nebraska Academy of Sciences, Lincoln, NE, April 25. http://puffin.creighton.edu/NRCSE/NeAcad.html.

Ayala, F. 2006. *Darwin and Intelligent Design*. Fortress, Minneapolis, MN.

Bakewell, J. 2006. Judgement Day. Review of *The God Delusion*, by Richard Dawkins. *Guardian*, September 23.

Balmer, R. 2006. *Thy Kingdom Come. An Evangelical's Lament. How the Religious Right Distorts the Faith and Threatens America*. Basic Books, New York.

Bambach, R. K. 1983. Response to Creationism. Review of *Abusing Science. The Case against Creationism*, by P. Kitcher; *Creation and Evolution: Myth or Reality?* by N. Newell; *Science on Trial: The Case for Evolution*, by D. Futuyma; *Scientists Confront Creationism*, edited by L. R. Godfrey; *The Monkey Business: A Scientist Looks at Creationism*, by N. Eldredge; and *Christianity and the Age of the Earth*, by D. A. Young. *Science* 220:851–53.

Banerjee, N., and A. Berryman. 2006. At Churches Nationwide, Good Words for Evolution. *New York Times*, February 13.

Barbour, I. G. 1990. *Religion in an Age of Science*. Harper, San Francisco, CA.

———. 2000. *When Science Meets Religion: Enemies, Strangers, or Partners?* Harper, San Francisco, CA.

Barrow, J. D., and F. Tipler. 1986. *The Anthropic Cosmological Principle*. Oxford University Press, Oxford, UK.

Berger, P. L., J. Sacks, D. Martin, and T. Weiming, eds. 1999. *The Desecularization of the World: Resurgent Religion and World Politics*. Eerdmans, Grand Rapids, MI.

Bonhoeffer, D. 1972. *Letters and Papers from Prison*. E. Bethge, ed. MacMillan, New York.

Bowler, P. J. 2001. *Reconciling Science and Religion. The Debate in Early Twentieth-Century Britain*. University of Chicago Press, Chicago, IL.

Broad, W. J. 2006. The Oracle Suggests a Truce between Science and Religion. *New York Times*, February 28.

Brockman, J., ed. 2006. *Intelligent Thought: Science versus the Intelligent Design Movement*. Random House, New York.

Brooke, J., and G. Cantor. 1998. *Reconstructing Nature. The Engagement of Science and Religion*. Oxford University Press, Oxford, UK.

Brooke, J. H. 1991. *Science and Religion: Some Historical Perspectives*. Cambridge University Press, Cambridge, UK.

Bruce, S. 2002. *God Is Dead: Secularization in the West*. Blackwell, Oxford, UK.

Bryan, J. R. 1999. Science and Religion at the Turn of the Millennium. Pp. 1–16 in P. H. Kelley, J. R. Bryan, and T. A. Hansen, eds., *The Evolution-Creation Controversy II: Perspectives on Science, Religion, and Geological Education*, vol. 5. Paleontological Society Papers. Paleontological Society, Pittsburgh, PA.

Call, L. R. First Parish Church Unitarian Universalist of Northborough, MA. www.firstparishnorthboro.org/sermon.shtml?20080323.

Cartmill, M. 1998. Oppressed by Evolution. *Discover* 19(3): 78–83.

CBS News Poll. 2006. April 6–9. www.pollingreport.com/religion.htm.

Chadwick, O. 1993. *The Secularization of the European Mind in the Nineteenth Century*. Cambridge University Press, Cambridge, UK.

Chang, K. 2006. Few Biologists but Many Evangelicals Sign Anti-Evolution Petition. *New York Times*, February 21.

Chesworth, A., ed. 2002. *Darwin Day Collection One: The Single Best Idea Ever*. Tangled Bank, Albuquerque, NM.

Clergy Letter Project. www.butler.edu/clergyproject/clergy_project.htm.

Cohen, I. B., ed. 1990. *Puritanism and the Rise of Modern Science: The Merton Thesis*. Rutgers University Press, New Brunswick, NJ.

Collins, F. S. 2006. *The Language of God. A Scientist Presents Evidence for Belief*. Free Press, New York.

Conser, W. H., Jr. 1993. *God and the Natural World: Religion and Science in Antebellum America*. University of South Carolina Press, Columbia.

Conway Morris, S. 2003. *Life's Solution: Inevitable Humans in a Lonely Universe*. Cambridge University Press, Cambridge, UK.

Coyne, J. A. 2000. Is NOMA a No Man's Land? Review of *Rocks of Ages: Science and Religion in the Fullness of Life*, by S. J. Gould. *Times Literary Supplement*, June 9.

Croce, P. J. 1995. *Science and Religion in the Era of William James. Eclipse of Certainty, 1820–1880*. University of North Carolina Press, Chapel Hill.

Darwin, C. 1993. Letter from Charles Darwin to Asa Gray, May 22, 1860. Pp. 223–26 in F. Burkhardt, J. Browne, D. M. Porter, and M. Richmond, eds., *The Correspondence of Charles Darwin*, vol. 8. Cambridge University Press, Cambridge, UK.

Davis, E. B., and R. Collins. 2000. Scientific Naturalism. Pp. 201–7 in G. B. Ferngren, ed., *The History of Science and Religion in the Western Tradition: An Encyclopedia*. Garland, New York.

Dawkins, R. 1986. *The Blind Watchmaker*. Norton, New York.

———. 1995. *River out of Eden. A Darwinian View of Life*. Basic Books, New York.

———. 2003. *A Devil's Chaplain. Reflections on Hope, Lies, Science, and Love*. Houghton Mifflin, Boston, MA.

———. 2006. *The God Delusion*. Houghton Mifflin, Boston, MA.

Dean, C. 2006. Faith, Reason, God and Other Imponderables. *New York Times*, July 25.

DeMaio, T. J. 1984. Social Desirability and Survey Measurement: A Review. Pp. 257–82 in C. F. Turner and E. Martin, eds., *Surveying Subjective Phenomena*, vol. 2, Russell Sage Foundation, New York.

Dembski, W. A. 2004. *The Design Revolution: Answering the Toughest Questions about Intelligent Design*. InterVarsity, Downers Grove, IL.

Dennett, D. C. 1995. *Darwin's Dangerous Idea: Evolution and the Meanings of Life*. Simon and Schuster, New York.

———. 2006. *Breaking the Spell: Religion as a Natural Phenomenon*. Viking, New York.

Dickerson, R. E. 1990. Letter to a Creationist: Seeking the Middle Ground. *Science Teacher*, September, pp. 49–53.

Discovery Institute. 2003. "The Wedge Document": So What? www.discovery.org/scripts/viewDB/filesDB_download.php?id=349. (Orig. pub. 2003.)

Dodson, E. O., and P. Dodson. 1985. *Evolution: Process and Product*. Van Nostrand Rinehold, New York.

Dodson, P. 1996. *The Horned Dinosaurs*. Princeton University Press, Princeton, NJ.

———. 1997. God and the Dinosaurs. *American Paleontologist* 5(2): 6–8.

———. 1999. Faith of a Paleontologist. Pp. 183–95 in P. H. Kelley, J. R. Bryan, and T. A. Hansen, eds., *The Evolution-Creation Controversy II: Perspectives on Science, Religion, and Geological Education*, vol. 5. Paleontological Society Papers. Paleontological Society, Pittsburgh, PA.

———. 2004. Cosmos and Creation: A Scientist Deepens His Faith While Defending It. *Catholic Standard and Times*, June 24, p. 1.

———. 2006. How Intelligent Is Intelligent Design? *American Paleontologist* 14(1): 23–26.

Domning, D. 2006. *Original Selfishness: Original Sin and Evil in the Light of Evolution.* Ashgate, Burlington, VT.

Draper, J. W. 1874. *History of the Conflict between Religion and Science.* Appleton, New York.

Drees, W. B. 1996. *Religion, Science and Naturalism.* Cambridge University Press, Cambridge, UK.

———. 2003. Naturalism. Pp. 593–97 in J. W. V. van Huyssteen, ed., *Encyclopedia of Science and Religion*, vol. 2. Macmillan Reference USA/Thomson-Gale, New York.

Edgell, P. 2006. Interview on NPR. *On the Media.* December 15. www.onthemedia.org.

Edgell, P., J. Gerteis, and D. Hartmann. 2006. Atheists as "Other": Moral Boundaries and Cultural Membership in American Society. *American Sociological Review* 72(2): 211–34.

Edis, T. 2003. A World Designed by God: Science and Creationism in Contemporary Islam. Pp. 117–25 in P. Kurtz, ed., *Science and Religion: Are They Compatible?* Prometheus Books, Amherst, NY.

Eisenberg, C. 2006. The Evolution of a Compromise: In Effort to Challenge. *Newsday*, February 12.

Emerson, R. W. 1841. Self-Reliance. Repr., 1980, pp. 25–52 in *The Collected Works of Ralph Waldo Emerson: Essays; First Series*, vol. 2. Harvard University Press, Cambridge, MA.

Ferngren, G. B., ed. 2002. *Science and Religion. A Historical Introduction.* Johns Hopkins University Press, Baltimore, MD.

Fingarette, H. 1969. Self-Deception. Routledge, Kegan, and Paul, London. Repr., 2000, with new chapter. University of California Press, Berkeley.

———. 1998. Self-Deception Needs No Explaining. *Philosophical Quarterly* 48(192): 289–301.

Flanagan, O. 2006. Varieties of Naturalism. Pp. 430–52 in P. Clayton and Z. Simpson, eds., *The Oxford Handbook of Religion and Science*. Oxford University Press, Oxford, UK.

Forrest, B. 2000. Methodological and Philosophical Naturalism: Clarifying the Connection. *Philo* 3:7–29.

Forrest, B., and P. R. Gross. 2004. *Creationism's Trojan Horse: The Wedge of Intelligent Design*. Oxford University Press, Oxford, UK.

Fowler, R. B., A. D. Hertzke, L. R. Olson, and K. R. den Dulk. 2004. *Religion and Politics in America: Faith, Culture, and Strategic Choices*, 3rd ed. Westview, Boulder, CO.

Gingerich, O. 2006. *God's Universe*. Belknap, Harvard University Press, Cambridge, MA.

Godfrey, S. J., and C. R. Smith. 2005. *Paradigms on Pilgrimage. Creationism, Paleontology, and Biblical Interpretation*. Clements, Toronto, ON.

Goodenough, U. 1998. *The Sacred Depths of Nature*. Oxford University Press, Oxford, UK.

———. 1999. The Holes in Gould's Semipermeable Membrane between Science and Religion. Review of *Rocks of Ages: Science and Religion in the Fullness of Life*, by S. J. Gould. *American Scientist* 87(3): 264–69.

Gould, S. J. 1983. *Hen's Teeth and Horse's Toes*. Norton, New York.

———. 1994. The Persistently Flat Earth. *Natural History* 103(3): 12–19. Repr., 1995, The Late Birth of a Flat Earth. Pp. 38–50 in S. J. Gould, *Dinosaur in a Haystack: Reflections in Natural History*. Harmony Books, New York.

———. 1995. *Dinosaur in a Haystack: Reflections in Natural History*. Harmony Books, New York.

———. 1997. Non-Overlapping Magisteria. *Natural History* 106(3): 16–22, 60–62. Repr., 1998, pp. 269–84 in S. J. Gould, *Leonardo's Mountain of Clams and the Diet of Worms: Essays on Natural History*. Harmony Books, New York.

———. 1999. *Rocks of Ages: Science and Religion in the Fullness of Life*. Ballantine, New York.

———. 2001. *I Have Landed*. Harmony Books, New York.

———. 2002. *The Structure of Evolutionary Theory*. Harvard University Press, Cambridge, MA.

Graffin, G. W. 2003. Monism, Atheism, and the Naturalist World-View: Perspectives from Evolutionary Biology. Ph.D. diss., Cornell University, Ithaca, NY.

Graffin, G. W., and W. B. Provine. 2007. Evolution, Religion and Free Will. *American Scientist* 95(4): 294–97.

Grant, E. 2001. *God and Reason in the Middle Ages*. Cambridge University Press, Cambridge, UK.

Griffin, D. R. 2000. *Religion and Scientific Naturalism: Overcoming the Conflicts*. State University of New York Press, Albany.

Guiderdoni, B. 2003. Islam: Contemporary Issues in Science and Philosophy. Pp. 465–69 in J. W. V. van Huyssteen, ed., *Encyclopedia of Science and Religion*, vol. 2. Macmillan Reference USA/Thomson-Gale, New York.

Harris, S. 2004. The End of Faith: Religion, Terror, and the Future of Reason. Norton, New York.

———. 2006. *Letter to a Christian Nation*. Knopf, New York.

Harris Poll. 1998. No. 41, August 12. www.harrisinteractive.com.

———. 2006. No. 80, October 31. www.harrisinteractive.com.

Harrison, P. 1998. *The Bible, Protestantism, and the Rise of Natural Science*. Cambridge University Press, Cambridge, UK.

Hasker, W. 1986. The Foundations of Theism: Scoring the Quinn-Plantinga Debate. *Faith and Philosophy* 15(1): 60–67.

Haught, J. F. 1995. *Science and Religion: From Conflict to Conversation*. Paulist, New York.

———. 2000. *God after Darwin: A Theology of Evolution*. Westview, Boulder, CO.

———. 2003. *Deeper Than Darwin: The Prospect for Religion in the Age of Evolution*. Westview, Boulder, CO.

———. 2006. God and Evolution. Pp. 697–712 in P. Clayton and Z. Simpson, eds., *The Oxford Handbook of Religion and Science*. Oxford University Press, Oxford, UK.

Hazen, R. 2002. Why Should You Be Scientifically Literate? www.actionbioscience.org/newfrontiers/hazen.html.

Heilbron, J. L. 1999. *The Sun in the Church: Cathedrals as Solar Observatories*. Harvard University Press, Cambridge, MA.

Hitchens, C. 2007. *God Is Not Great. How Religion Poisons Everything*. Hatchette Book Group, Twelve, New York.

Hollinger, D. A. 1989. Justification by Verification: The Scientific Challenge to the Moral Authority of Christianity in Modern America. Pp. 116–35 in M. J. Lacey, ed., *Religion and Twentieth-Century American Intellectual Life*. Cambridge University Press, Cambridge, UK.

Holt, J. 2006. Beyond Belief. Review of *The God Delusion*, by Richard Dawkins. *New York Times Book Review*, October 22, pp. 1, 11–12.

Hooykaas, R. 1972. Religion and the Rise of Modern Science. Eerdmans, Grand Rapids, MI.

International Theological Commission. 2004. Communion and Stewardship: Human Persons Created in the Image of God. The July 2004 Vatican Statement on Creation and Evolution. www.bringyou.to/apologetics/p80.htm.

Jacoby, S. 2004. *Freethinkers. A History of American Secularism.* Metropolitan Books/ Henry Holt, New York.

Jammer, M. 2003. Materialism. Pp. 538–43 in J. W. V. van Huyssteen, ed., *Encyclopedia of Science and Religion,* vol. 2. Macmillan Reference USA/Thomson-Gale, New York.

John Paul II. 1996. Truth Cannot Contradict Truth (Magisteria). Address to the Pontifical Academy of Sciences, October 22. *L'Osservatore Romano* 30:3, 7. Repr., 1997, *Quarterly Review of Biology* 72:381–83. www.nsceweb.org/resources/articles/3992_statements_from_religious_orgs_12_19_2002.asp#rom2.

Johnson, G. 2005. Agreeing Only to Disagree on God's Place in Science. *New York Times,* Sept. 27

———. 2006a. A Free-for-All on Science and Religion. *New York Times,* November 21. Repr., 2007, *Skeptical Inquirer* 31(2): 24–26.

———. 2006b. Scientists on Religion. Review of *God's Universe,* by Owen Gingerich; *The Language of God,* by Francis Collins; *The God Delusion,* by Richard Dawkins; and *The Varieties of Scientific Experience,* by Carl Sagan. *Scientific American* 295(4): 94–95.

Johnson, G. S. 1998. *Did Darwin Get It Right? Catholics and the Theory of Evolution.* Our Sunday Visitor, Huntington, IN.

Johnson, P. E. 1995. *Reason in the Balance: The Case against Naturalism in Science, Law and Education.* InterVarsity, Downers Grove, IL.

———. 2000. *The Wedge of Truth: Splitting the Foundations of Naturalism.* InterVarsity, Downers Grove, IL.

Jones, J. E., III. 2005. Memorandum Opinion. *Kitzmiller v. Dover Area School District.* Case 4:04-cv-02688 JEJ. Document 342. U.S. District Court for the Middle District of Pennsylvania, December 20. www.pamd.uscourts.gov/kitzmiller/kitzmiller_342.pdf.

Kazin, M. 2006. *A Godly Hero. The Life of William Jennings Bryan.* Knopf, New York.

Kelley, P. H. 1983. Evolutionary Patterns of Eight Chesapeake Group Molluscs: Evidence for the Model of Punctuated Equilibria. *Journal of Paleontology* 57: 581–98.

———. 1984. Multivariate Analysis of Evolutionary Patterns of Seven Miocene Chesapeake Group Molluscs. *Journal of Paleontology* 58:1235–50.

———. 1999. An Honors Course on Evolution and Creationism: Teaching Experiences in the Deep South. Pp. 217–25 in P. H. Kelley, J. R. Bryan, and T. A. Hansen, eds., *The Evolution-Creation Controversy II: Perspectives on Science, Religion, and Geological Education*, vol. 5. Paleontological Society Papers. Paleontological Society, Pittsburgh, PA.

———. 2000. Studying Evolution and Keeping the Faith. *Geotimes* 45(12): 22–23, 41.

———. 2009. Stephen Jay Gould's Winnowing Fork: Science, Religion, and Creationism. Pp. 171–88 in W. D. Allmon, P. H. Kelley, and R. M. Ross, eds., *Stephen Jay Gould: Reflections on His View of Life*. Oxford University Press, New York.

Kelley, P. H., and T. A. Hansen. 1993. Evolution of the Naticid Gastropod Predator-Prey System: An Evaluation of the Hypothesis of Escalation. *Palaios* 8:358–75.

———. 2003. The Fossil Record of Drilling Predation on Bivalves and Gastropods. Pp. 113–39 in P. H. Kelley, M. Kowalewski, and T. A. Hansen, eds., *Predator-Prey Interactions in the Fossil Record*. Kluwer Academic/Plenum, New York.

Kitcher, P. 2007. *Living with Darwin. Evolution, Design, and the Future of Faith*. Oxford University Press, Oxford, UK.

Kloppenberg, J. T. 1986. *Uncertain Victory: Social Democracy and Progressivism in European and American Thought, 1870–1920*. Oxford University Press, Oxford, UK.

Knight, C. C. 2001. *Wrestling with the Divine: Religion, Science, and Revelation*. Fortress, Minneapolis, MN.

Krauss, L. M. 2006. Sermons and Straw Men. Review of *The God Delusion*, by Richard Dawkins. *Nature* 443:914–15.

Kuhn, T. 1970. *The Structure of Scientific Revolutions*, 2nd ed. University of Chicago Press, Chicago, IL.

LaBute, N. 2007. *Wrecks and Other Plays*. Faber and Faber, New York.

Lacey, M. J., ed. 1989. *Religion and Twentieth-Century American Intellectual Life*. Cambridge University Press, Cambridge, UK.

Lachman, S. P. 1993. *One Nation under God: Religion in Contemporary American Society*. Harmony Books, New York.

Lakoff, G. 2002. *Moral Politics: How Liberals and Conservatives Think*, 2nd ed. University of Chicago Press, Chicago, IL.

Larson, E. J., and L. Witham. 1997. Scientists Are Still Keeping the Faith. *Nature* 386:435–36.

———. 1998. Leading Scientists Still Reject God. *Nature* 394:313.

———. 1999. Scientists and Religion in America. *Scientific American,* September, 88–93.

Lazar, A. 1999. Deceiving Oneself or Self-Deceived? On the Formation of Beliefs "Under the Influence." *Mind* 108:265–90.

Lerner, M. 2006. *The Left Hand of God: Taking Back Our Country from the Religious Right.* Harper, San Francisco, CA.

Leuba, J. H. 1916. *Belief in God and Immortality: A Psychological, Anthropological and Statistical Study.* Sherman, French, Boston, MA.

Lindberg, D. C. 1992. *The Beginnings of Western Science: The European Scientific Tradition in Philosophical, Religious, and Institutional Context, 600 BC to AD 1450.* University of Chicago Press, Chicago, IL.

Lindberg, D. C., and R. L. Numbers. 2003. Introduction. Pp. 1–5 in D. Lindberg and R. Numbers, eds., *When Science and Christianity Meet.* University of Chicago Press, Chicago, IL.

Livingstone, D. 1984. *Darwin's Forgotten Defenders: The Encounter between Evangelical Theology and Evolutionary Thought.* Regent College, http://regent.gospelcom.net/rcp/aboutrcp.html.

Marsden, G. M. 1996. *Soul of the American University: From Protestant Establishment to Established Nonbelief.* Oxford University Press, Oxford, UK.

Marsden, G. M., and B. J. Longfield, eds. 1992. *The Secularization of the Academy.* Oxford University Press, Oxford, UK.

Marty, M. E. 2006. Religiosecular Meditations: Sightings. *Marty Center at the University of Chicago Divinity School.* http://marty-center.uchicago.edu/sightings/archive_2006/0619.shtml.

Matsumura, M. 1996. *Voices for Evolution,* 2nd ed. National Center for Science Education, Berkeley, CA.

McGrath, A. 1998. *Science and Religion: An Introduction.* Blackwell, Oxford, UK.

———. 2004. *The Twilight of Atheism: The Rise and Fall of Disbelief in the Modern World.* Doubleday, New York.

McLaughlin, B. P., and A. O. Rorty, eds. 1988. *Perspectives on Self-Deception.* University of California Press, Berkeley.

Merton, R. K. 1938. Science, Technology, and Society in Seventeenth-Century England. *Osiris* 4:360–362.

Meyers, R. 2006. *Why the Christian Right Is Wrong: A Minister's Manifesto for Taking Back Your Faith, Your Flag, Your Future.* Jossey-Bass, San Francisco, CA.

Miller, A. I. 2004. Presentation of the Paleontological Society Medal to Richard K. Bambach. *Journal of Paleontology* 78(4): 815–16.

Miller, J. B., ed. 2001. *An Evolving Dialogue. Theological and Scientific Perspectives on Evolution.* Trinity Press International, Harrisburg, PA.

Miller, K. B., ed. 2003. *Perspectives on an Evolving Creation.* Eerdmans, Grand Rapids, MI.

Miller, K. R. 1999. *Finding Darwin's God: A Scientist's Search for Common Ground between God and Evolution.* Harper, New York.

———. 2005. Looking for God in All the Wrong Places. Answering the Religious Challenge to Evolution, Pp. 13–21 in J. Cracraft and R. W. Bybee, eds., *Evolutionary Science and Society: Educating a New Generation.* BSCS, Colorado Springs, CO. Revised proceedings of the BSCS, AIBS Symposium, Chicago, IL, November 2004.

Mooney, C. 2005. *The Republican War on Science.* Basic Books, New York.

Moore, J. A. 1993. *Science as a Way of Knowing.* Harvard University Press, Cambridge, MA.

Moore, J. R. 1979. *The Post-Darwinian Controversies.* Cambridge University Press, Cambridge, UK.

Moreland, J. P. 1989. *Christianity and the Nature of Science.* Baker Book House, Grand Rapids, MI.

Myers, P. Z. 2006. Bad Religion. Review of *The God Delusion,* by Richard Dawkins. *Seed* 2(7): 88–89.

Nasr, S. H. 2006. Islam and Science. Pp. 71–86 in P. Clayton and Z. S. Simpson, eds., *The Oxford Handbook of Religion and Science.* Oxford University Press, Oxford, UK.

National Academy of Sciences. 1998. *Teaching about Evolution and the Nature of Science.* National Academy Press, Washington, DC.

National Science Board. 2006. *Science and Engineering Indicators 2006.* National Science Foundation, Arlington, VA. www.nsf.gov/statistics/seind06/.

Nebraska Religious Coalition for Science Education. http://puffin.creighton.edu/nrcse.

Newsweek Poll. 2000. Princeton Survey Research Associates, April 13–14. www.pollingreport.com/religion2.htm.

Nielsen, K. 1993. No! A Defense of Atheism. Pp. 48–56 in J. P. Moreland and K. Nielsen, eds., *Does God Exist? The Debate between Theists and Atheists.* Prometheus Books, Amherst, NY.

Numbers, R. L. 2003. Science without God: Natural Laws and Christian Beliefs. Pp. 265–86 in D. C. Lindberg and R. L. Numbers, eds., *When Science and Christianity Meet*. University of Chicago Press, Chicago, IL.

———. 2006. *The Creationists: From Scientific Creationism to Intelligent Design*, exp. ed. Harvard University Press, Cambridge, MA.

Obama, B. 2006. *The Audacity of Hope. Thoughts on Reclaiming the American Dream*. Crown, New York.

Olson, R. 2004. *Science and Religion, 1450–1900: From Copernicus to Darwin*. Johns Hopkins University Press, Baltimore, MD.

Orr, H. A. 1999. Gould on God: Can Religion and Science Be Happily Reconciled? Review of *Rocks of Ages: Science and Religion in the Fullness of Life*, by S. J. Gould. *Boston Review*, October–November, pp. 24–26.

———. 2006. A Mission to Convert. Review of *The God Delusion*, by Richard Dawkins; *Six Impossible Things before Breakfast: The Evolutionary Origins of Belief*, by Lewis Wolpert; and *Evolution and Christian Faith: Reflections of an Evolutionary Biologist*, by Joan Roughgarden. *New York Review of Books*, January 11, pp. 21–24.

Overbye, D. 2005. Philosophers Notwithstanding, Kansas School Board Redefines Science, November 15. www.nytimes.com/2005/11/15/science/sciencespecial2/15evol.html.

Pennock, R. 1999. *Tower of Babel: The Evidence against the New Creationism*. MIT Press, Cambridge, MA.

Perakh, M. 2004. *Unintelligent Design*. Prometheus Books, Amherst, NY.

Peterson, M., W. Hasker, B. Reichenbach, and D. Basinger. 2003. *Reason and Religious Belief: An Introduction to the Philosophy of Religion*, 3rd ed. Oxford University Press, Oxford, UK.

Petto, A. J., and L. R. Godfrey, eds. 2007. *Scientists Confront Intelligent Design and Creationism*. Norton, New York.

Pigliucci, M. 1999. Gould's Separate "Magisteria": Two Views. Review of *Rocks of Ages: Science and Religion in the Fullness of Life*, by S. J. Gould. *Skeptical Inquirer* 23(6): 53–56.

———. 2000. Personal Gods, Deism, and the Limits of Skepticism. *Skeptic* 8(2): 38–45.

———. 2002. *Denying Evolution: Creationism, Scientism, and the Nature of Science*. Sinauer Associates, Sunderland, MA.

Polkinghorne, J. 1998. *Belief in God in an Age of Science*. Yale University Press, New Haven, CT.

———. 2005. *Exploring Reality. The Intertwining of Science and Religion*. Yale University Press, New Haven, CT.

Provine, W. B. 1987. A Review of E. J. Larson, Trial and Error: The American Controversy over Creation and Evolution. *Academe* 73:50–52.

———. 1988a. Evolution and the Foundation of Ethics. *MBL Science* 3:25–29.

———. 1988b. Progress in Evolution and Meaning of Life. Pp. 49–74 in M. H. Nitecki, ed., *Evolutionary Progress*. University of Chicago Press, Chicago, IL.

———. 1988c. Scientists, Face It! Science and Religion Are Incompatible. *Scientist*, September 5, p. 10.

———. 2006. Evolution, Religion, and Science. Pp. 667–80 in P. Clayton and Z. Simpson, eds., *The Oxford Handbook of Religion and Science*. Oxford University Press, Oxford, UK.

Raymo, C. 1998. *Skeptics and True Believers: The Exhilarating Connection between Science and Religion*. Walker, New York.

———. 2005. *Honey from Stone: A Naturalist's Search for God*. Cowley, Lanham, MD.

Roberts, J. H. 1988. *Darwinism and the Divine in America. Protestant Intellectuals and Organic Evolution, 1859–1900*. University of Wisconsin Press, Madison.

Roberts, J. H., and J. Turner. 2000. *The Sacred and the Secular University*. Princeton University Press, Princeton, NJ.

Rolston, H., III. 1999. *Genes, Genesis, and God: Values and Their Origins in Natural and Human History*. Cambridge University Press, Cambridge, UK.

———. 2006. *Science and Religion. A Critical Survey*. Templeton Foundation, Philadelphia, PA.

Roughgarden, J. 2006. *Evolution and Christian Faith: Reflections of an Evolutionary Biologist*. Island, Washington, DC.

Ruse, M. 1997. Review of *Rocks of Ages: Science and Religion in the Fullness of Life*, by S. J. Gould. *Metanexus Magazine*. www.metanexus.net/metanexus_online/show_article2.asp?id = 3044

———. 2001a. *Can a Darwinian Be a Christian? The Relationship between Science and Religion*. Cambridge University Press, Cambridge, UK.

———. 2001b. *The Evolution Wars. A Guide to the Debates*, rev. ed. Rutgers University Press, New Brunswick, NJ.

———. 2005. *The Creation-Evolution Struggle*. Harvard University Press, Cambridge, MA.

———. 2007. Fighting the Fundamentalists. Chamberlain or Churchill? *Skeptical Inquirer* 31(2): 38–41.

Russell, R. J. 1998. Special Providence and Genetic Mutation: A New Defense of Theistic Evolution. Pp. 191–223 in R. J. Russell, W. R. Stoeger, and F. J. Ayala, eds., *Evolution and Molecular Biology: Scientific Perspectives on Divine Action*. Vatican Observatory Publications, Vatican City.

———. 2006. Quantum Physics and the Theology of Non-interventionist Objective Divine Action. Pp. 579–95 in P. Clayton and Z. Simpson, eds., *The Oxford Handbook of Religion and Science*. Oxford University Press, Oxford, UK.

Scopes, J. T. 1925. *The World's Most Famous Court Trial. Tennessee Evolution Case*. National Book Company, Cincinnati, OH.

Scott, E. 1996. Dealing with Antievolutionism. Pp. 15–28 in J. Scotchmoor and F. K. McKinney, eds., *Learning from the Fossil Record*, vol. 2. Paleontological Society Papers. Paleontological Society, Pittsburgh, PA.

———. 1997. Antievolutionism and Creationism in the United States. *Annual Review of Anthropology* 26:263–89.

———. 1999. But I Don't Believe in Evolution! The Science Teacher's Dilemma. *Journal of Religion and Education* 26(2): 67–75.

———. 2004. *Evolution and Creationism. An Introduction*. Greenwood, Westport, CT.

Scott-Kakures, D. 1996. Self-Deception and Internal Irrationality. *Philosophy and Phenomenological Research* 56(1): 31–56.

Shanks, N. 2004. *God, the Devil, and Darwin. A Critique of Intelligent Design Theory*. Oxford University Press, Oxford, UK.

Shermer, M. 2002. *Why Do People Believe Weird Things?* Rev. and exp. ed. Henry Holt, New York.

———. 2006. *Why Darwin Matters: The Case against Intelligent Design*. Times Books, New York.

Silk, M. 1988. *Spiritual Politics: Religion and America since World War II*. Simon and Schuster, New York.

Simpson, G. G. 1978. *Concession to the Improbable: An Unconventional Autobiography*. Yale University Press, New Haven, CT.

Smith, W. C. 1962. *The Meaning and End of Religion. A New Approach to the Religious Traditions of Mankind*. Macmillan, New York.

Stark, R. 1999. Secularization, R. I. P. *Sociology of Religion* 60(3): 249–73.

Stenger, V. J. 2007. God: The Failed Hypothesis; How Science Shows That God Does Not Exist. Prometheus Books, Amherst, NY.

Taverne, D. 2005. *The March of Unreason. Science, Democracy, and the New Fundamentalism*. Oxford University Press, Oxford, UK.

Thomson, K. S. 2005. *Before Darwin: Reconciling God and Nature*. Yale University Press, New Haven, CT.

Trivers, R. 1985. *Social Evolution.* Benjamin Cummings, Menlo Park, CA.

Turner, J. 1985. *Without God, Without Creed: The Origins of Unbelief in America.* Johns Hopkins University Press, Baltimore, MD.

Tyson, N. D. 2001. Holy Wars: An Astrophysicist Ponders the God Question. *Skeptical Inquirer* 25(5): 10–14.

United Church of Christ. 1992. *Creationism, the Church, and the Public School.* Board for Homeland Ministries. www.ncseweb.org/resources/articles/5025_statements_from_religious_orgs_12_19_2002.asp#home.

United Presbyterian Church in the U.S.A. 2002. 214 General Assembly of the Presbyterian Church. www.ncseweb.org/resources/articles/3992_statements_from_religious_org_12_19_2002.asp#pres3.

Wallis, J. 2005. *God's Politics: Why the Right Gets It Wrong and the Left Doesn't Get It.* Harper, San Francisco, CA.

Watson, R. A. 2000. Three Biologists and Religion. Review of *The Sacred Depths of Nature,* by Ursula Goodenough; *Reason for Hope: A Spiritual Journey,* by Jane Goodall; and *Rocks of Ages: Science and Religion in the Fullness of Life,* by Stephen J. Gould. *Quarterly Review of Biology* 75(2): 159–65.

Weinberg, S. 1993. Dreams of a Final Theory: The Search for the Ultimate Laws of Nature. Pantheon, New York.

———. 1999. A Designer Universe? *New York Review of Books,* October 21. Repr., 2003, pp. 31–40 in P. Kurtz, ed., *Science and Religion. Are They Compatible?* Prometheus Books, Amherst, NY.

———. 2007. A Deadly Certitude. Review of *The God Delusion,* by Richard Dawkins. *Times Literary Supplement,* January 19, pp. 5–6.

White, A. D. 1896. A History of the Warfare of Science with Theology in Christendom, 2 vols. Appleton, New York.

Wilson, D. B. 2000. The Historiography of Science and Religion. Pp. 3–11 in G. B. Ferngren, ed., *The History of Science and Religion in the Western Tradition: An Encyclopedia.* Garland, New York.

Witham, L. 2003. *By Design: Science and the Search for God.* Encounter Books, San Francisco, CA.

———. 2005. *The Measure of God: Our Century-Long Struggle to Reconcile Science and Religion.* Harper, San Francisco, CA.

Wolf, G. 2006. The New Atheism: The Church of the Non-Believers. *Wired,* November, pp. 182–93.

Young, M., and T. Edis, eds. 2004. *Why Intelligent Design Fails: A Scientific Critique of the New Creationism.* Rutgers University Press, New Brunswick, NJ.

SELECTED RESOURCES RELEVANT TO INTELLIGENT DESIGN

BOOKS

Comfort, N. 2007. *The Panda's Black Box: Opening Up the Intelligent Design Controversy*. Johns Hopkins University Press, Baltimore, MD. A collection of essays that address the source of the intelligent design controversy and investigate the various factors that continue to form it. Contributors include law professor and historian Edward Larson and historian of science Daniel Kevles. The book engages readers interested in comprehending the important implications of the debate over evolution.

Falk, D. R. 2004. *Coming to Peace with Science: Bridging the Worlds between Faith and Biology*. InterVarsity Press, Downers Grove, IL. A well-informed and personal examination of evolution by a biologist and committed evangelical Christian who is a professor at Point Loma Nazarene University. The book contains an extensive section on the fossil record.

Forrest, B., and P. R. Gross. 2004. *Creationism's Trojan Horse: The Wedge of Intelligent Design*. Oxford University Press, New York. Philosopher Barbara Forrest and biologist Paul Gross scrutinize the goals and strategies of the intelligent design movement as articulated in the Discovery Institute's famous Wedge Document. They consider the religious foundations of ID, the political ambitions of ID proponents, and the lack of a scientific ID hypothesis.

Gould, S. J. 1999. *Rocks of Ages: Science and Religion in the Fullness of Life*. Ballantine Books, New York. In this book, published a few years before his death, paleontologist and historian of science Steve Gould argues that religion and science are separate domains of human thought and experience. Referring to these realms as "nonoverlapping magisteria," he offers this principle as a simple and conventional resolution to the supposed conflict between science and religion.

Hedman, M. 2007. *The Age of Everything: How Science Explores the Past.* University of Chicago Press, Chicago, IL. Covering a wide range of time scales, astronomer Matthew Hedman illustrates how scientists have determined the ages of various objects and events, from the colonization of the great pyramids of Egypt to the origins of the 14-billion-year-old universe.

Humes, E. 2007. *Monkey Girl: Evolution, Education, Religion and the Battle for America's Soul.* Ecco, New York. Journalist Edward Humes provides a detailed and thorough account of the events in Dover, Pennsylvania, that gave rise to the *Kitzmiller v. Dover* court case.

Johnson, K. 2007. *Cruisin' the Fossil Freeway: An Epoch Tale of a Scientist and an Artist on the Ultimate 5,000-Mile Paleo Road Trip.* Fulcrum, Golden, CO. Paleontologist and chief curator at the Denver Museum of Nature and Science chronicles his drive with artist Ray Troll across the American West in search of fossils. Throughout their journey to numerous remote locations, they demonstrate the ubiquity of fossils as they encounter suburban T. rexes, killer Eocene pigs, ancient fossilized forests, and paleo-seashores.

Kelley, P. H., J. R. Bryan, and T. A. Hansen, eds. 1999. *The Evolution-Creation Controversy II: Perspectives on Science, Religion, and Geological Education,* vol. 5. Paleontological Society Papers, Paleontological Society, Pittsburgh, PA. First presented as a short course for the Paleontological Society, this work primarily deals with creation science and Flood geology arguments. With its respectful approach to religious belief in general, this book is especially useful to teachers.

Larson, E. J. 2001. *Evolution's Workshop: God and Science on the Galápagos Islands.* Basic Books, New York. A historical discussion of both the scientific and theological controversies focused on "evolution's workshop"—the Galápagos Islands. This book details, in their cultural, social, and intellectual contexts, Darwin's explorations on the islands; the fight for control of research on the Galápagos; and contemporary efforts by creationists to use the Galápagos in their own teachings.

Livingstone, D. N. 1987. *Darwin's Forgotten Defenders: The Encounter between Evangelical Theology and Evolutionary Thought.* Eerdmans, Grand Rapids, MI. An excellent review, from the mid-nineteenth century to the present, of the response of evangelical scientists and theologians to Darwin's theory. It reveals that evolution by natural selection was (and is) accepted, sometimes enthusiastically, by a number of evangelical scholars.

Miller, K. B., ed. 2003. *Perspectives on an Evolving Creation.* Eerdmans, Grand Rapids, MI. Written almost entirely by devout evangelical scientists, this edited volume of twenty contributions addresses evolution from within a variety of disciplines and in the context of orthodox Christian faith. The book includes substantial Earth science content, as well as a direct response from a physicist and chemist to the ID argument.

Miller, K. R. 1999. *Finding Darwin's God: A Scientist's Search for Common Ground.* Harper Collins, New York. Using examples from molecular biology, physics, astronomy, and geology, Catholic believer and biologist Kenneth Miller describes the difference between evolution as "validated scientific fact" and as a developing theory. The book constitutes an excellent rebuttal to young-Earth creationists, intelligent design advocates, and atheistic materialists.

National Academy of Sciences. 2008. *Science, Evolution and Creationism.* National Academies Press, Washington, DC. Written for those wanting to learn more about the science of evolution, this revised edition of *Science and Creationism* offers a succinct overview of recent advances from the fossil record, molecular biology, and evolutionary developmental biology that have yielded significant, new, and convincing evidence for evolution. The book will be important reading for educators involved in the classroom teaching of evolution.

Numbers, R. L. 2006. *The Creationists: From Scientific Creationism to Intelligent Design.* Harvard University Press, Cambridge, MA. First published in 1992, this authoritative account of the roots of creationism has been expanded and updated. Ronald Numbers, a historian of science, chronicles the rise of Flood geology, the assertion of present-day scientific creationists that most fossils date back to Noah's Flood and its aftermath.

Pennock, R. T. 1999. *Tower of Babel: The Evidence against the New Creationism.* MIT Press, Cambridge, MA. By analogy with the evolution of languages, philosopher Robert Pennock critiques intelligent design arguments and compares views of new creationists with those of the old.

Petto, A. J., and L. R. Godfrey, eds. 2007. *Scientists Confront Intelligent Design and Creationism.* Norton, New York. The sixteen essays in this collection include at least two that will be very useful to Earth science educators: Victor Stenger's "Physics, Cosmology, and the New Creationism"; and Brent Dalrymple's "The Ages of the Earth, Solar System, Galaxy, and Universe." Of particular use to teachers, students, and general readers are essays that discuss approaches to educational and legal policies that can help to preserve the integrity of public and scientific institutions.

Prothero, D. 2007. *Evolution: What the Fossils Say and Why It Matters.* Columbia University Press, New York. Paleontologist Donald Prothero details the transitional forms in the fossil record and engages subjects germane to them, including Flood geology and the radiometric dating of rocks. The book is especially valuable because it clarifies the nature and value of fossil evidence for evolution.

Rudwick, M. J. S. 1985. *The Meaning of Fossils: Episodes in the History of Paleontology*, 2nd ed. University of Chicago Press, Chicago, IL. Historian of science Martin Rudwick provides an excellent, readable account of important ideas about Earth history. The book helps the reader understand the way in which the present view of the history of the Earth developed.

Scotchmoor, J., D. Springer, B. Breithaupt, and T. Fiorillo. 2002. *Dinosaurs: The Science behind the Stories*. American Geological Institute, Alexandria, VA. Essays in this book emphasize the principles of science that guide paleontological research and interpretation. The papers by Springer and Scotchmoor ("Unearthing the Past") and Sampson ("The Science of Paleontology: New View on Ancient Bones") are especially useful for teachers.

Scott, E. C., and G. Branch. 2006. *Not in Our Classrooms: Why Intelligent Design Is Wrong for Our Schools*. Beacon, Boston, MA. Written by the executive director and the deputy director of the National Center for Science Education, respectively, the book is an understandable and nuanced critique of intelligent design. By describing the history of the intelligent design movement, its invalid claims, and the pedagogical, legal, and religious problems it poses, the book equips readers to argue against teaching intelligent design as science.

Strahler, A. N. 1999. *Science and Earth History: The Evolution/Creation Controversy*. Prometheus Books, Buffalo, NY. Geomorphologist Arthur Strahler scrutinizes evidence from astronomy and planetary science to zoology in order to demonstrate that the name *creation science* is in fact an oxymoron. His investigation is well reasoned and reveals "creation science" to be a system of belief.

Young, M. 2004. *Why Intelligent Design Fails: A Scientific Critique of the New Creationism*. Rutgers University Press, New Brunswick, NJ. Biologists, computer scientists, physicists, archaeologists, and mathematicians examine information-based arguments for intelligent design, as well as the claim of irreducible complexity, and show that they do not challenge Darwinian evolution.

Zimmer, C. 1998. *At the Water's Edge*. Touchstone, New York. Carl Zimmer provides an account that reveals examples of transitional forms in the fossil record where the first tetrapods and walking whales were discovered.

WEB SITES

American Geological Institute. Geoscience Education Homepage. www.agiweb.org/education/index.html. The American Geological Institute is a nonprofit federation of forty-four geoscientific and professional associations that represent more than one hundred thousand geologists, geophysicists, and other Earth scientists. Founded in 1948, AGI provides information services to geoscientists, serves as a voice of shared interests in the profession, plays a major role in strengthening geoscience education, and strives to increase public awareness of the vital role the geosciences play in society's use of resources and interaction with the environment. The Geoscience Education Homepage has links to information on curriculum materials, professional development, and careers in the geosciences.

The Complete Work of Charles Darwin Online. http://darwin-online.org.uk/. A "virtual bookshelf" of the works of Charles Darwin, this Web site contains examples of all known publications of Charles Darwin. It includes more than forty thousand pages of searchable text and more than one hundred thousand electronic images. It does not cover Darwin's unpublished letters, which are the focus of the Darwin Correspondence Project.

Geological Society of America. *The Teaching of Evolution*. www.geosociety.org/positions/index.htm. This Web site includes a position statement on the teaching of evolution from the Geological Society of America, an organization of thousands of Earth scientists around the world committed to studying the Earth and sharing scientific findings.

National Center for Science Education. www.natcenscied.org/default.asp. The National Center for Science Education is a nationally recognized clearinghouse for information and advice related to keeping the teaching of evolution in the science classroom and creationism out of it. NCSE is the only national organization specializing in this issue.

Paleontological Research Institution and the Museum of the Earth. www.museumoftheearth.org. The Museum of the Earth contains eight thousand square feet of permanent exhibits, telling the history of life on Earth through the geological record of the northeastern United States. Unique elements include the skeletons of the Hyde Park Mastodon and Right Whale #2030 and the 544-square-foot mural *Rock of Ages Sands of Time*.

United States Geological Survey. *The USGS and Science Education*. http://education.usgs.gov/. The U.S. Geological Survey provides scientific information used to describe and understand the Earth; minimize loss of life and property from natural disasters; manage water and biological, energy, and mineral resources; and enhance and protect quality of life. The USGS helps educate the public about natural resources, natural hazards, geospatial data, and issues that affect humans and other living things. Online resources include lessons, data, and maps.

University of California Museum of Paleontology. www.ucmp.berkeley.edu/. The UCMP Web site contains exhibit sections on understanding evolution, geologic time, and the history of life through time. Also on the Web site are the paleontology portal, a monthly mystery fossil, and K-12 resources.

MEDIA

ABC News. 2006. *Nightline: Intelligent Design vs. Evolution.* ABC News correspondent Chris Bury reports on the campaign to bring intelligent design into the teaching of evolution by natural selection.

American Geological Institute. 2007. *Faces of Earth.* Science Channel. A four-part series that takes viewers on a trip with geoscientists through the history and the interconnected systems of the planet. With topics including the formation of the planet to the emergence of life, as well as the dynamic external and internal processes that have shaped the Earth, the program helps viewers understand how humans are both a force of nature and a product of our world.

Nova, 2007. *Judgement Day: Intelligent Design on Trial.* WGBH, Boston. This program details the significant federal case of *Kitzmiller v. Dover,* a battle over teaching evolution in public schools and the first legal test of intelligent design as a scientific theory. The film features trial reenactments based on court transcripts and interviews with key participants, including expert scientists and Dover parents, teachers, and town officials.

Olson, Randy. 2006. *Flock of Dodos: The Evolution–Intelligent Design Circus.* A comic and controversial feature documentary in which filmmaker, scientist, surfer, and evolutionary biologist Randy Olson looks at the intelligent design versus evolution controversy in Pennsylvania and Kansas.

ABOUT THE CONTRIBUTORS

WARREN D. ALLMON is the director of the Paleontological Research Institution (PRI) in Ithaca, NY, and the Hunter R. Rawlings III Professor of Paleontology in the Department of Earth and Atmospheric Sciences at Cornell University. He earned his A.B. in Earth sciences from Dartmouth College in 1982, and his Ph.D. in Earth and atmospheric sciences from Harvard University in 1988. Since 1992, he has been instrumental in rejuvenating PRI's internationally known fossil collections; in starting its local, regional, and national programs in Earth science education; and in planning and fundraising for the Museum of the Earth, PRI's $11 million education and exhibit facility that opened in September 2003. Allmon is a fellow of the Geological Society of America and the recipient of the 2004 Award for Outstanding Contribution to Public Understanding of Geoscience from the American Geological Institute.

DAVID W. GOLDSMITH is associate professor of geology and paleontology at Westminster College in Salt Lake City, Utah, where he teaches paleontology as well as the history and philosophy of science. He received his undergraduate degree from Colgate and his Ph.D. from Harvard University. His paleontological research focuses on the morphology and ecology of Molluscs. He is also interested in scientific epistemology and the history of evolutionary theory. Presently he is studying the distribution of Recent gastropods on the shores of Bear Lake on the Utah-Idaho border.

TIMOTHY H. HEATON is professor of geology and chair of the Department of Earth Sciences and Physics at the University of South Dakota. He earned his bachelor's degree from Brigham Young University and his Ph.D. in Earth and atmospheric sciences from Harvard University. His research projects include investigations of ice age vertebrates in southeast Alaska, Quaternary paleontology of the Great Basin, and Tertiary rodents of western North America.

PATRICIA H. KELLEY is professor of geology at the University of North Carolina, Wilmington. She received her undergraduate degree from College of Wooster and her Ph.D. from Harvard University. Her interests include the tempo and mode of evolution and the role of biological factors such as predation in evolution, which she investigates using coastal plain mollusc fossils. Dr. Kelley has authored or coauthored over forty refereed papers and has edited three books, including one on evolution and creationism. She is a former president of the Paleontological Society. In 2003 she received the Outstanding Educator Award from the Association for Women Geoscientists. She is currently a distinguished speaker on the topic of evolution and creationism for the Paleontological Society and National Association of Geoscience Teachers.

KEITH B. MILLER is research assistant professor of geology at Kansas State University. He earned his B.A. in geology from Franklin and Marshall College and his Ph.D. from the University of Rochester. Miller is editor of *Perspectives on an Evolving Creation* (2003, Eerdmans, Grand Rapids, MI), an anthology of essays on evolution by Christian scholars. He is also a board member of the Kansas Citizens for Science, a not-for-profit educational organization that promotes a better understanding of science.

CHARLES E. MITCHELL is SUNY distinguished teaching professor at SUNY, Buffalo. He earned his Ph.D. in geology from Harvard University in 1983. Mitchell studies graptolites from the Ordovician Period, and he is motivated by a desire to understand the evolutionary processes that have formed the world in which we live and that have given shape to its history. He approaches his subject by testing specific hypotheses using quantitative means whenever possible but also keeps sight of the unique properties of biological and geological phenomena.

DONALD R. PROTHERO is professor of geology at Occidental College in Los Angeles, and lecturer in geobiology at the California Institute of Technology in Pasadena. He earned his M.A., M.Phil., and Ph.D. degrees in geological sciences from Columbia University in 1982, and a B.A. in geology and biology from the University of California, Riverside. He is currently the author, coauthor, editor, or coeditor of twenty-one books and almost two hundred scientific papers, including five leading geology textbooks and three trade books, as well as edited symposium volumes and other technical works. He is on the editorial board of *Skeptic* magazine and in the past has served as an associate or technical editor for *Geology*, *Paleobiology*, and *Journal of Paleontology*. He is a fellow of the Geological Society of America, the Paleontological Society, and the Linnaean Society of London and has also received fellowships from the Guggenheim Foundation and the National Science Foundation. He has been the vice president of the Pacific Section of SEPM (Society of Sedimentary Geology) and has served five years as the program chair for the Society of Vertebrate Paleontology. In 1991, he received the Schuchert Award of the Paleontological Society for being an outstanding paleontologist under the age of forty. He has also been featured on several television documentaries, including episodes of *Paleoworld* and *Walking with Prehistoric Beasts*.

JILL S. SCHNEIDERMAN is professor of Earth science at Vassar College. She earned her bachelor's degree in geology from Yale University and her Ph.D. in Earth and atmospheric sciences from Harvard University in 1987. Jill worked as a congressional science fellow in the office of former Democratic leader of the U.S. Senate Tom Daschle during the 104th Congress, as a postdoctoral fellow at the Smithsonian's National Museum of Natural History, and as a Fulbright fellow at the University of the West Indies in Trinidad. She is the editor of and contributor to *The Earth around Us: Maintaining a Liveable Planet* (2003, Westview, Boulder, CO). She is a fellow of the Geological Society of America. Her research interests include provenance of heavy minerals in detrital sediments in recent deltas, gender and water resources, and the intersection of Earth science, feminism, and environmental justice.

MARK TERRY is chair of the Science Department of the Northwest School in Seattle, which he cofounded in 1978 and where he served as head of school from 1983 to 1990. Mark studied philosophy and history at Pomona College, earned his B.A. in physical anthropology from the University of Washington, and his M.A.T. in Science Education from Cornell University, where he was a Shell Merit Fifth Year Scholar. Mark has taught evolutionary science in public and independent schools for over thirty years in New York, California, Oregon, and Washington and was a keynote speaker at the National Conference on Teaching Evolution in Berkeley in 2000. He is author of *Teaching for Survival* (1971, Ballantine, New York) and the first *Environmental Education Guidelines for Washington State Secondary Schools* (1976, OSPI, Olympia), as well as other articles on environmental and science education. Mr. Terry is a member of Phi Beta Kappa, Phi Delta Kappa, and the Society of Vertebrate Paleontology.

ALLISON R. TUMARKIN-DERATZIAN is a teaching assistant professor in the Department of Earth and Environmental Science at Temple University. She earned her Ph.D. from the University of Pennsylvania. Dr. Tumarkin-Deratzian's research primarily focuses on bone growth in modern and fossil tetrapods, using both gross examination and histological study in thin-section. Other projects include investigations of ontogeny and evolution in ceratopsian dinosaurs, and study of crocodyliforms from the Cretaceous of Egypt (as part of the Bahariya Dinosaur Project). Tumarkin-Deratzian teaches courses in physical geology, evolution and extinctions, stratigraphy, and paleontology.

INDEX

Abrahamic faiths. *See* Christianity; Islam; Judaism

accommodationism, 182, 197–219; concepts of the God of, 215–19; general explanations for, 203–5; among nonscientists, 197–203; among scientists, 205–15

Adam and Eve story, 26, 28, 29, 173

Addison, Thomas, 82

aepycamelines, 47

Africa: rift zones of, 16, 65; transitional fossil forms in, 47, 52

African Methodist Episcopal Church, 201

Agassiz, Louis, 148

Alabama, fossil beds in, 48

Aleutian Islands, earthquakes in, 153

Algeria, fossil beds in, 52

Allmon, Warren D., 1–7, 180–227, 247

Alpine metamorphic rocks, 6

Alvarez, L. W., 167

alvarezsaurids, 71–72

Ambulocetus natans, 51, 54

amebelodonts, 52

American Jewish Congress, 176, 200

American Philosophical Association, 124

amniotes, 43

amphibians, 43

anancines, 52

Anaximander, 142

Anglican Church, 4

anthropic principle, 4, 190

apologetics, religious, 95–96

Appalachian Mountains, 15; garnets of, 17, *18*

Aquinas, Thomas, 151

archaeocetes, 48–52

Archaeopteryx, 39, 60–61, 64–73

Arif, M., 51

Aristotle, 96, 98, 143, 151, 152

Arizona State University, 200

Arkansas Board of Education, 1, 164, 169, 207

artiodactyls, 42, 48, 49

asteroid impact hypothesis, 167

astronomy, 27–30, 78, 223

atheism, 117, 166, 181–82, 193–94, 213, 219, 222; creationist equation of evolutionary science with, 121–22, 126, 199–200; extreme, 109, 110, 191–93, 197, 202, 216–17, 220; methodological naturalism attacked as promoting, 118–22, 136; spirituality and, 113n5; theistic science and, 125–27
Atkins, Peter, 191
Atlantic Ocean, 16
atomic theory, 214
Augustine, 197
Austin, S. A., 24, 25
Australia, detrital zircons in gneisses in, 5
Ayala, Francisco, 203–4

Babuna, Oktar, 113n4
Babylonian exile, 173–74, 206
Bacon, Francis, 143–44, 149; *Novum Organum*, 143, 145
Bactrian camels, 47
Bambach, Richard, 210–12
baraminology, 42–43, 48
Barbour, Ian G., 181, 182, 206
Barrow, J. D., 190
Basilosaurus, 49–50
Baumgardner, J. R., 24, 25
Baylor Science and Research Project, 137
Beagle (ship), 83
bears, 54
Behe, Michael, 100, 170; *Darwin's Black Box*, 94–95, 134, 135, 151–52
Bible, 22, 24, 25, 55, 56, 87, 100, 169, 172–73, 188, 199, 209, 225n5; Genesis, 21, 22, 24, 26–28, 112, 169, 173–75, 197; Isaiah, 218; Jeremiah, 206–7; Job, 174–75; Proverbs, 56, 174; Psalms, 174–75; Revelation, 86–87
Big Bang, 27, 203, 210
Bighorn Basin (Wyoming) fossil beds, 45
Big Tent, 78, 79
Biological Society of Washington, *Proceedings* of, 40, 87
biology, 11, 12, 22, 28–30, 39, 78, 79, 42, 223; Cambrian, 32; complexity of, 31; evolutionary, 4, 136, 180, 191, 195, 200, 208; transformational, 142
birds, evolution from dinosaurs of, 40, 59–61; intelligent design arguments against, 61–73
Bonaparte, Napoleon, 192
Brightman, Edgar, 124–25
Britain: accommodation of science and religion in, 198; intellectual tradition in, 143–45
Broad, William, 201–2
brontotheres, 45
Brown University, 207
Bryan, William Jennings, 198, 225n5
Buckland, William, 83, 84, 87; *The Bridgewater Treatises*, 82
Buddhism, 112, 113n5
Buffon, Georges, 144
Burnet, Thomas, 144
Bush, A., 24
Bush, George W., 54

Caldwell, Billy R.: *Geology in the Bible*, 2
California, fossils found in, 47
Call, Lon Ray, 216
Calvert, John, 118–20
Calvin, John, 197–98
Cambrian Explosion, 32–33, 39, 40
camels, evolution of, 42, 47–48
Canada, fossil beds of, 168
Carroll, Lewis: *Through the Looking Glass*, 42
catastrophic plate techtonics, 13, 14, 16
Caudipteryx, 60
causation, 96–100; Aristotelian notion of, 151; divine, Roman Catholic doctrine of, 217; in logic of Intelligent Design, 154; in methodological naturalism, 117; supernatural, 166, 170

CBS News polls, 194
Cenozoic era, 24, 47, 102
Central Conference of American Rabbis, 200
Chambers, Robert, 101
chemistry, 11, 195, 223; cause-and-effect laws of, 119
Chesapeake Bay, fossils in sediments of, 163
Chesterton, G. K., 180
China: feathered dinosaur fossils found in, 60, 61, 64, 66, 71; volcanic islands rimming coast of, 15
Christianity, 21, 93, 95, 111, 127, 170, 171, 183, 194, 225n2; accommodation of science and, 198, 201, 206, 207, 210, 215–16, 221; belief in irreconcilability of evolutionary theory and, 121; fundamentalist, 39, 55–56, 78, 87, 113n4, 182, 224; liberal to moderate, 189; methodological naturalism and, 123; nonphysical soul in, 133; progressive creationist, 26–28; purpose in, 99; socially conservative, 195; Theistic Evolutionist, 22
cladistic methodology, 61–62, 69, 72; cladograms, 62, *63*, 64–65, 104, *105*, *106;* naming conventions in, 69–70
clams, 163
Clepysaurus, 16
Clergy Letter Project, 201
climate change, 2
collateral relatives, 64
Collins, Francis, 206
components, law of, 17
Compsognathus, 60, 65, 71
Confuciusornis, 60
Congress, U.S., 213
Conser, W. H., Jr., 225n1
Constitution, U.S., 54, 203
Cornell University, 206
cosmogonists, 144
creationism, 2, 94, 152, 164, 168–71, 180–82, 209; "appearance of age" view in, 217; consistency conundrum in, 195, 197; denominations opposing teaching of, 200–201; in God Spectrum, 194, 195; Islam and, 113n4; methodological naturalism attacked by advocates of, 117–19, 121, 131, 135–36; old-Earth, 5, 12, 21, 26–28; perspectives on geology in, 21–34; political and social agenda of, 6; progressive, 21, 22, 26–31, 33; theistic science as objective of, 126–29. *See also* intelligent design (ID); young-Earth creationism
Creation Museum (Petersburg, Kentucky), 2
Creation Research Society, 169
creation science, 121–23, 164, 169, 172, 206
Creation Science Research Center, 169
Cretaceous Period, 60, 64–66, 72, 107, 167
cross-cutting relationships, law of, 17
cross-sections, geological, 14–17

Daeschler, E. B., 167–68
Dalanistes, 51
Darrow, Clarence, 225n5
Darwin, Charles, 4–6, 23, 88–89, 111, 152, 163, 190; "branching bush" model of, 40; celebration of birthday of, 201; deductive method of, 145–51; earthquake studies of, 80, 82–84; *On the Origin of Species by Means of Natural Selection*, 21, 43, 65, 67, 94, 99–100, 141–51, 156, 157, 198
Darwin, Erasmus, 101, 142
Darwinism, 55, 111, 182, 198, 206, 223; extreme atheist view of, 109, 110, 191, 192, 202; Roman Catholic acceptance of, 201; theistic evolution and, 126;

Darwinism *(continued)*
use as pejorative term, 121; Wedge strategy against, 79
Davis, P., 41, 43; *Of Pandas and People*, 39, 51, 88
Dawkins, Richard, 101, 110, 188, 190, 191, 199, 202, 222
Day-Age Theory, 21, 28
Dean, Cornelia, 202
death, introduction into world of, 26, 28
deinotheres, 52
deism, 206
Dembski, William, 11–12, 33, 56, 95, 122, 126–27, 129–30, 132, 137, 170; *The Design Revolution*, 11
Democratic Party, 220
Dennett, Daniel C., 188, 191, 196, 202, 222
derived characters, shared, 62, 65, 72
design. *See* intelligent design (ID)
design filter, 132–34
Devonian Period, 103
de Vries, Paul, 123–24, 126, 139n17
diacodexeids, 42, 48
Diceratherium, 47
dichobunids, 42, 48
Dickerson, R. E., 190
Dinosauria, 67
dinosaurs, 16, 23; debate over extinction of, 166–67; evolution of birds from, 40, 59–61
directional evolution, 102–9
discoverability, 84
Discovery Institute, 54, 77–80, 85, 157–58; Center for Science and Culture (CSC), 3, 78, 84, 86, 87
Disney Studios, 175
District Court, U.S., 1
divine action, attempts to detect, 127–29
Dobson, James, 78
Dobzhansky, Theodosius, 206
Dodson, Peter, 208–9, 220
dolphins, 48, 51, 62
Domning, Daryl, 207
Dover (Pennsylvania) School District, 2, 55, 88, 110, 120, 165, 170, 171, 199, 207
Draper, John William: *History of Conflict between Religion and Science*, 123
dromedary, 47

earthquakes, 80–85, 153, 170
eclipses, solar, 85
ecosystems, diversity in, 102
Edwards v. Aguillard (1987), 170
Egypt, fossil beds in, 49 52
Eldredge, N., 41
elephants, evolution of, 52–54
Elser, Jim, 200
Emerson, Ralph Waldo, 222
enaliarctines, 54
Enlightenment, 14, 79–80, 82, 83, 89, 206
Enuma elish (Babylonian creation epic), 173
Eocene epoch, 43–45, *46*, 47–49, 51, 53
Eohippus, 43
Episcopal Church, 176, 200, 201
erratics, glacial, 147–48
evangelicals, 123, 182, 204
evolution, 7, 23, 32, 39, 120, 121, 126, 195, 223–24; accommodation of religion and, 89, 164, 176, 199–202, 206, 209, 214, 215, 219–21; branching bush model of, 40–41, 44; deductive approach to study of, 146; gradualist view of, 41–42; incompatibility of Linnean system with, 69; prediction and hypothesis testing in study of, 167–68; progressive creationist rejection of, 28, 30; role of metaphysics in debate on, 93–113; theistic, 22, 126. *See also* fossil record; human evolution; natural selection; Wedge Document and strategy
Expelled. No Intelligence Allowed (movie), 79

extinctions, 23, 28–31, 41, 47, 48, 52, 101; of dinosaurs, debate over, 166–67; Linnean hierarchy and, 67; mass, 102, 106, 107
extraterrestrial impact. *See* asteroid impact hypothesis

falsifiability, principle of, 153, 166, 168, 169; accommodationism and, 220–21
Falwell, Jerry, 87
Fawcett, Henry, 150
Fayûm fossil beds (Egypt), 52
First Conference on Creation Geology (Cedarville, Ohio, 2007), 2
fish, 62, 65, 70, 167; lobe-finned, 168
Fisher, R. A., 206
Fitzgerald, F. Scott, 180
Flood, biblical, 22, 24–26, 89
Focus on the Family, 78, 85
Fontenelle, Bernard le Bovier de: *Conversations on the Plurality of Worlds*, 4
Fordham gneiss, 15, 16
fossil record, 15, 17, 19, 23, 28, 29, 31, 39–40; attribution to Flood of, 24; cladistic methodology for, 61–62; directionality in, 102–9; escalation in, 209; as evidence of death, 26; of molluscs, 163; progressivist views and, 100–102; rapid diversification in, 32; transitional forms in, 40–56, 59–73, 167–68
fossils: in Africa, 47, 52; in Alabama, 48; in Algeria, 52; in Canada, 168; in Egypt, 52; in Mongolia, 44; in Morocco, 53; in Texas, 47; in Utah, 47; in Wyoming, 45
French intellectualism, 143
Froede, Carle: *Geology by Design*, 2, 12
Futuyma, Douglas, 200

Gallup polls, 2
Gap Theory, 21
garnets, 17–18, *18*
Gaviocetus, 51
genetics, Mendelian, 79
Geological Society of America, 86
Geological Society of London, 144–45; *Transactions* of, 145
Germany: *Archaeopteryx* fossil found in, 60; nineteenth-century, science in, 145, 148
Gigantocamelus, 47
Gingerich, Owen, 207
Gingerich, Philip, 49
giraffes, 47, 54
Gishlick, A. D., 59
glaciation, 25, 29, 165, 166
Godfrey, Stephen, 207
God of the Gaps argument, 154, 172, 218
God Spectrum, 184–87, 189–219; complete naturalism in, 190–94; complete supernaturalism in, 194–97; paleontologists in, 207–12. *See also* accommodationism
Goldsmith, David, 6, 141–59, 247
Gonzalez, Guillermo, 84–85; *The Privileged Planet*, 79, 84–86, 88
Gordon, Bruce, 137
Gould, Stephen Jay, 4, 40, 41, 101, 102, 164, 187–89, 209
Grand Canyon, 12–13, 89, 164–66
graptolites, 103–8, *105*, *106*
Graves, Dan: *The Earth Will Reel from Its Place*, 2
Gray, Asa, 210
"great chain of being," 40–41
Greeks, ancient, 142
Greenland, detrital zircons in gneisses in, 5
Greenspahn, F. E., 174, 175
guanacos, 47

Harris, Sam, 191–92
Harris polls, 194, 225n3, n4
Harvard University, 40, 207, 210
Haugh, John, 188–89, 216–18

Heaton, Timothy, 6, 21–38, 248
Hebrew Bible, translations of, 172–73
hippos, 48, 49
Hirmantian Mass Extinction (HME), 106
Hobbes, Thomas, 217
Homogalax, 44, 45
Homo sapiens, 65
Hooker, J. D., 146
Hooykaas, R., 225n1
horses, evolution of, 39, 42–47, *46*, 48, 54
Hudson River, 6; geological cross-section of, 14–17, *15*
human evolution, 23, 93–95, 110–12; ancient Greek concept of, 142; branching bush model of, 40–41; cladistics and, 65, 70; irreducible complexity argument and, 94; progressivist view of, 100–102, 108; in old-earth-creationism, 28–30
Human Genome Project, 207
Humboldt, Alexander von, 148
Humphreys, D. Russell, 24
Hussain, S. T., 51
Hutton, James, 5, 13
Huxley, Thomas Henry, 5, 43
hydrocarbons, fossil, 28
Hyers, Conrad, 173, 185
hyperevolution, 26
Hyracotherium, 43

ice ages, 25, 29, 47, 147–48, 165, 166
Illustra Media, 85
inclusion trails, *18*, 18–19
inerrancy, 27
information theory, 95, 170
Institute for Creation Research, 169, 199
intelligent design (ID), 1, 6, 7, 11, 31–34, 89, 93–94, 110, 164, 180, 182; absence of theory in, 136–37; anthropic principle and, 4; apologetics of, 95–96, 109; commitment to purpose in, 94–95; discoverability argument of, 84–85; evolution of, 168–71; false dichotomy in, 87; federal court decision against, 2, 55, 120, 165, 170, 171, 199; fossil record misrepresented in, 39–42, 51, 54; in God Spectrum, 194–97; historical roots of, 79–80, 88; intellectual context of, 155–57; irreducible complexity argument in, 5, 12–15, 59, 94, 99, 170–72; logical structure of, 151–55; methodological naturalism attacked by advocates of, 117–23, 129–37; political agenda of, 54–56, 77–79; progressive creationists and, 21–22, 31, 34; scientists quoted out of context in, 41, 44; theistic science as objective of, 126–29, 171; theological and educational dangers of, 171–72; weakness of assertions for scientific legitimacy of, 141–42, 157–58
Intelligent Design Network, 118, 119
interfaith community, scientific inquiry in, 125
InterVarsity press, 78
Inwood limestone, 15, 16
iridium, 167
irreducible complexity, 5, 12–15, 59, 94, 99, 170–72
Islam, 113n5, 176, 183, 225n2; fundamentalist, 113n4, 224

Jamaica, 53
James, William, 198
Japan, volcanic islands of, 15
Jenkins, F. A., 167–68
Jeremiah, 206
Jesus, 175, 208, 219
John Paul II, Pope, 176, 201, 216
Johnson, Phillip E., 39, 56, 120, 122, 126, 131, 170, 195; *Darwin on Trial*, 120–21
Jones, John E., III, 2, 54, 120, 164–65, 170, 199
Journal of Geoscience Education, 86

Judaism, 176, 183, 200, 225n2; ancient, 173–75, 197, 206–7
Judeo-Christian tradition, 54, 113n5, 191. *See also* Christianity; Judaism
Jurassic Period, 60, 64–66, 72

Kansas State Board of Education, 88, 118–20, 122, 123, 137n1, 156–57, 196–97
Kazin, M., 198
Kelley, Patricia H., 7, 163–79, 209–10, 248
kenosis, 215–16
Kenyon, D., 41, 43; *Of Pandas and People*, 39, 51, 88
Kepler, Johannes, 12
Kitzmiller v. Dover Area School District (2005), 2, 55, 88, 110, 120, 165, 170, 171, 199, 207
Kosnik, M., 102
Krauss, L. M., 193
Kubrick, Stanley, 85

Labutte, Neil: *Wrecks*, 180
"ladder of life," 40
Lamarck, Jean Baptiste, 101, 142
Laplace, Pierre-Simon, 192
Larson, E. J., 193
"layer cake geology," 13
Liberty University, 87–88
Lidgard, S., 102
Linnean classification system, 61, 66–70
Lisbon earthquake (1755), 80–83
llamas, 47
lunar rocks, dating of, 23
Lutherans, 176
Lyell, Charles, 4–5, 80–84, 87, 89

mammals, 70; shared derived characters of, 62; transitional forms of, 40–56. *See also* amebelodonts; archaeocetes; artiodactyls; bears; brontotheres; camels, evolution of; deinotheres; elephants, evolution of; giraffes; guanacos; hippos; horses, evolution of; llamas; mammoths; manatees, evolution of; mastodonts; mesonychids; oromerycids; perissodactyls; phenacodontids; pinnipeds; *Poebrotherium;* rhinos, evolution of; seals; Sirenia; tapirs; ungulates; vicuñas; whales
mammoths, 52
manatees, evolution of, 53–54
Manhattan schist, 15, 16
Marsh, O. C., 43
Marty, M. E., 182
Masoretic scholars, 173
mass extinction. *See* extinctions
mastodonts, 52, 53
Matsumura, A., 176
McLean v. Arkansas Board of Education (1981), 1, 164, 169, 171, 207
McLuhan, Marshall, 86
meaning, 96–100; self in perspective to, 111
Mendel, Gregor, 79
Mesohippus, 44
mesonychids, 49, 51
Mesopotamian creation stories, 173–75
Mesozoic era, 15, 24, 64, 66, 102
metamorphic rocks, 18; deformed, 17–19; folded, 14
metaphysical naturalism, 124, 125
metaphysics, 132; in debate on evolution, 93–113
meteorites, dating of, 23
Methodist Church, 176, 200
methodological naturalism (MD), 98, 117–37, 165, 183, 201, 207; attack on, 118–23, 196; "intelligence" and design filter and, 132–35; origin and meaning of, 123–32; as restriction on search for "truth," 129–32
Meyer, Stephen C., 32–33, 87

Microraptor, 60
Mill, John Stuart, 150, 155, 156; *A System of Logic*, 149–50
Miller, Keith B., 6, 117–40, 207, 248
Miller, Kenneth R., 207, 215, 224
minerals, 5, 17–18
Miocene epoch, 47
Miohippus, 44
miolabines, 47
miracles, 33, 155, 186, 189, 197, 221, 222; accommodationist views on, 205–8, 210, 213, 215, 219; cause-and-effect processes as plausible account of, 128; in progressive creationism, 29, 31; in young-Earth creationism, 26
"missing links," 41
Mitchell, Charles E., 6, 93–116, 248
Mivart, St. George, 94
Moeritherium, 52–53
molluscs, 163
Mongolia, fossil beds in, 44
Moon, Sun-Myung, 56
Moore, A. L., 111–12
Morocco, fossil beds of, 53
Morris, Henry, 24, 121–22; *The Genesis Flood*, 22
Morris, Simon Conway, 207
Moses, 197
mountains, formation of, 18
Mount Saint Helens, 2
multicultural community, scientific inquiry in, 125
Murray, M. J., 24
Muslims. *See* Islam

National Academy of Sciences (NAS), 193, 213, 214
National Association of Geoscience Teachers, 87
National Museum of Natural History, 210
National Religious Broadcasters, 56
naturalism. *See* metaphysical naturalism; methodological naturalism; philosophical naturalism
natural selection, 2, 32, 79, 93, 113n3, 141–42, 223; accommodation of religion and, 191, 195, 203, 208, 214; deductive argument for, 147, 152, 155; genetic variations as basis of, 217; inversion of cause and effect in, 99–100; irreducible complexity argument against, 59; Lyellian uniformitarianism and, 4; in metaphysical naturalism, 110–12; progressive creationist assertion of ineffectiveness of, 27, 29
natural theology, 79, 80–83, 88, 206
Nebraska Religious Coalition for Science Education, 201
Nelson, Paul, 86, 129–31, 136–37
neo-Darwinian theory, 32
Neptunists, 144
Newark Basin, 14; sandstones and shales of, 16
Newton, Isaac, 143, 148, 149, 155; *Principia*, 143–44
Newsweek polls, 194, 225n4
New York City, folded metamorphic rocks of, 14
New York Times, The, 86, 201–2
Noah, 22, 24–26, 42, 88
non-overlapping magisteria (NOMA), 187–89
Numbers, Ronald L., 21, 139n17
Numidotherium, 52–53

Oard, M. J., 24, 25
Obama, Barack, 204, 220
Oligocene epoch, 47, 52
ontology, 215; natural, 217
Ordovician Period, 103, 104, 106, 107
Origen, 173, 175
oromerycids, 42, 48
Overton, William R., 1, 164
Owen, Richard, 67

Pakicetus, 49
Pakistan, 49, 51
Palaeomastodon, 52
Paleocene epoch, 53, 167
Paleontological Society, 209
Paleozoic era, 24, 32, 102
Paley, William, 81–82, 84, 88–89, 151, 155; *Natural Theology*, 4, 81
Palisades, 14; trap ridge of, 16
Pangea, 16, 84
paradox of the soil, 5
Parker, Gary, 121–22
Pennock, R., 192, 195–96, 199–200, 202, 217
perissodactyls, 42, 45, *46*
Permian mass extinction, 102, 107
petroleum, 28
Pezosiren portelli, 53–54
phenacodontids, 45
philosophical naturalism, 119–21, 183
Phiomia, 52
Phosphatherium, 53
photomicrographs, 17
phylogenic systematics. *See* cladistic methodology
physics, 11, 23; of flight, 107; laws of, 96, 119
Pigliucci, M., 190
pinnipeds, 54
Plantiga, Alvin, 95, 100
plate tectonics, 25, 79, 153, 170, 214
Pleistocene epoch, 25
Pliocene epoch, 47
Plutonists, 144
Poebrodon, 47, 48
Poebrotherium, 47
Polkinghorne, John, 207
porphyroblasts, 18
predator-prey coevolution, 209
Presbyterian Church, 164, 176, 200, 201, 209, 210
Price, George McCready, 22, 24
primitive traits, shared, 62
Proboscidea, 52, 53
Proceedings of the Biological Society of Washington, 40, 87
progress, notions of, 100–102; directionality and, 102–9
Progressive Creationists, 21
Protestants, 22, 113n4, 176, 198
Prothero, Donald, 6, 39–58, 249–49
protolabines, 47
Protorohippus, 43, 44
Provine, William B., 101, 191, 202, 213, 216–17
public education, 55, 77–79, 199–200; science literacy in, 135–37; strategy for including intelligent design in, 3, 55, 78–79, 85–89, 194. *See also* Arkansas Board of Education; Dover (Pennsylvania) School District; Kansas State Board of Education
Punch, 148
punctuated equilibrium, 41, 209
purpose, 100, 119; intelligent design commitment to, 94–95, 99; of inanimate objects, 97–99; self in perspective to, 111

quantum indeterminacy, 215, 217
Quarterly Review, The, 147

Radinskya, 44–45
Radioisotopes and the Age of Earth (RATE), 25–26
radiometric dating, 23, 25, 32
Raup, David, 41, 101, 102
Reasons to Believe, 27, 78
Regnery Publishing, 78, 85
relativity, 214
reptiles, 16, 43, 67
Reynolds, John Marks, 129, 131
rhinos, evolution of, 42, 44, 45, *46*, 47, 54
Richards, Jay W., 84–85; *The Privileged Planet*, 79, 84–86, 88
Rift Valley (East Africa), 16, 65

Robinson, S. J., 24
rocks, 5, 14–16, 17–19, 89
Rodhocetus, 51
Roman Catholic Church, 176, 198, 201, 208, 217
Ross, Hugh, 22, 27–31
Ross, Marcus, 86–88
Roughgarden, Joan, 206
Royal Geological Society, 146
Royal Society, 207
Rudin, A. J., 172
Ruse, Michael, 196, 197, 207, 224
Rushmore, Mount, 12
Rutiodon, 16

San Andreas Fault, 153
sarcopterigians, 168
Scheven, J., 24
schist: Manhattan, 15, 16; Vishnu, 89
Schneiderman, Jill, 1–7, 11–20, 249
science literacy, 135–37, 172
Science Research Foundation, 113n4
scientific method, 164, 165, 169, 176; induction versus deduction in, 142–51, 155
scientism, 124
Scopes trial, 198
Scotland, nineteenth-century science in, 145
Scott, E., 190
Scottish Enlightenment, 14
seals, 54
search for extraterrestrial intelligence (SETI), 134
sections, geological, 14–17
Sedgwick, Adam, 146
sedimentary rocks, 22–23; attribution to Flood of, 24, 25
self-organizational mechanisms, 32
Shubin, N. H., 167–68
Silurian Period, 107
Simons, E. L., 49
Simpson, George Gaylord, 108–9, 219; *The Meaning of Evolution*, 100
Sinosauropteryx, 60, 65
Sirenia, 53–54
Sjogren, Jody, 119
Smith, B. H., 49
Smithsonian Institution, 210
snails, 163
social acceptability bias, 204–5, 213–14
soil, paradox of, 5
solar eclipses, total, 85
Solomon, King, 174
South Dakota, Big Badlands of, 47
Southern Baptist Church, 201
speciation, 28–31; of Cambrian Explosion, 32
Spectator, The, 146
Spinoza, Baruch, 217
Steffen, L., 171
stegotetrabelodonts, 52
Steinbeck, John: *Cannery Row*, 97–99
Steinmetz, D. C., 173, 175
stenomylines, 47
storm intensity, 29
supernatural agent, acts of. *See* miracles
Swiss Society of Natural Sciences, 148

Takracetus, 51
tapirs, 42, 44, 45
taxonomy, Linnean, 61
Teilhard de Chardin, Pierre, 206
Tennessee, caves of, 2
Terry, Mark, 6, 77–92, 249
tetrapods, 43, 70; evolution from sarcopterigians of, 168; shared derived characters of, 62, 65
Texas, fossils found in, 47
theism, 21, 22, 110, 111, 121, 152, 192; accommodation of science and, 206; attempt to develop science based on, 100, 126–29, 181; God of the Gaps argument in, 154; Wedge strategy and, 87

theropod dinosaurs, feathered, 60–62, 64, 66, 68, 69, 71–73
Thewissen, J. G. M., 51
Tipler, F., 190
Titanotylopus, 47
tolerance, propensity for, 203, 213
Toles, Tom, 100
transformational biology, 142
transformative processes, directionality of, 102–9
Triassic Period, 64
Tumarkin-Deratzian, Allison, 6, 59–74, 249–50
Tyson, Neil Degrasse, 213

unconformity, *15*, 16
ungulates, 48, 49, 54
Unification Church, 56
uniformitarianism, 4–5
Unitarian-Universalist Association, 176, 200, 216
United Church of Christ, 176, 200, 210
University of North Carolina, at Wilmington, 209
University of Pennsylvania, 208
University of Rhode Island, 86
Utah, fossils found in, 47

van Valen, Leigh, 49
Vardiman, L., 25
Vermont, Appalachian mountains of, 17, *18*
vicuñas, 47
Virginia Tech, 210
Vishnu schist, 89
volcanoes, 15, 28, 167
Voltaire: *Candide*, 80–81

Wagner, P. J., 102
warfare metaphor, 123, 181
"watchmaker" argument, 151
Wedge Document and strategy, 3, 55, 78–79, 85–89, 194
Wells, Jonathan, 32, 43, 51, 56, 72–73; *Icons of Evolution*, 39, 65, 70, 72; *The Politically Incorrect Guide to Darwinism and Intelligent Design*, 72–73
whales, 62; evolution of, 43, 48–52, *50*, 54, 73
Wheaton College, 123
Whewell, William, 155, 156; *The Philosophy of the Inductive Sciences, Founded Upon Their History*, 149, 150
Whitcomb, John, 24; *The Genesis Flood*, 22
White, Andrew Dickson: *A History of the Warfare of Science with Theology in Christendom*, 123
Wilberforce, Samuel, 147
Williams, G. C., 101
Wise, Kurt P., 24, 25, 40
Witham, L., 193
Witschey, Walter, 176
Witt, Jonathan, 84
Wood, T. C., 24
Wyoming, 45

young-Earth creationism, 1, 5, 12, 21–26, 34, 39, 131–32; explanation of fossil record by, 24, 40–42, 44, 61; intelligent design compared with, 31–33; legal failures of, 21; methodological naturalism attacked by, 118, 124, 136; progressive creationist critique of, 27–28, 30; Wedge strategy of, 86
Yucatán Peninsula, evidence of asteroid impact in, 167

Zimmer, Carl: *At the Water's Edge*, 51
Zimmerman, Michael, 201

Text: 10.25/14 Fournier
Display: Fournier
Compositor: SNP Best-set Typesetter
Indexer: Ruth Elwell
Printer and binder: Maple-Vail Book Manufacturing Group